Universe, Human Immortality and Future Human Evaluation

Universe, Human Immortality and Future Human Evaluation

Alexander Bolonkin
Dr. Sci., Professor, Former Senior Researcher of NASA
and Scientific Laboratories of the USA Air Forces
USA

AMSTERDAM • BOSTON • HEIDELBERG • LONDON • NEW YORK • OXFORD
PARIS • SAN DIEGO • SAN FRANCISCO • SINGAPORE • SYDNEY • TOKYO

ELSEVIER

Elsevier
32 Jamestown Road, London NW1 7BY
225 Wyman Street, Waltham, MA 02451, USA

First edition 2012

British Library Cataloguing-in-Publication Data
A catalogue record for this book is available from the British Library

Library of Congress Cataloging-in-Publication Data
A catalog record for this book is available from the Library of Congress

ISBN: 978-0-12-415801-6

For information on all Elsevier publications
visit our website at elsevierdirect.com

This book has been manufactured using Print On Demand technology. Each copy is produced to
order and is limited to black ink. The online version of this book will show color figures where
appropriate.

Contents

Preface

Immortality is the most cherished dream and the biggest wish of any person. In this book, the author shows that the problem of immortality can be solved only by changing the biological human into an artificial form. After natural death, man and his conscious mind may be converted into a new electronic form. Such an immortal person made of chips and super-solid material (the E-man or E-being as it was called in the author's earlier articles and book) will have incredible advantages compared to conventional biological people. An E-man will need no food, no dwelling, no air, no sleep, no rest, and no ecologically pure environment. His brain will work from radioisotopic batteries (which will work for decades) and muscles that will work on small nuclear engines. Such a being will be able to travel into space and walk on the sea floor with no aqualungs. He will change his face and figure. He will have super-human strength and communicate easily over long distances to gain vast amounts of knowledge in seconds (by rewriting his brain). His mental abilities and capacities will increase millions of times. It will be possible for such a person to travel huge distances at the speed of light. The information of one person like this could be transported to other planets with a laser beam and then placed in a new body. Biological people gradually will transfer into E-beings. The biological civilization little by little will convert to a higher electronic civilization. In further development of technology, the electronic civilization may be converted into an electromagnetic civilization and so on.

This is the popular book about the universe, the development of new technologies in the twenty-first century, and the future of the human race. The author shows that a human soul is only the information in a person's head. He offers a new unique method for rewriting the main brain information in chips without any damage to the human brain.

This is the scientific prediction of the nonbiological (electronic) civilization and immortality of human beings. Such a prognosis is predicated upon a new law, discovered by the author, for the development of complex systems. According to this law, every self-copying system tends to be more complex than the previous system, provided that all external conditions remain the same. The consequences are disastrous: humanity will be replaced by a new civilization created by intellectual robots (which the author refers to as "E-humans" and "E-beings"). These creatures, whose intellectual and mechanical abilities will far exceed those of man, will require neither food nor oxygen to sustain their existence. They may have emotions and be capable of developing science, technology, and their own intellectual abilities thousands of times faster than humans can; they will, in essence, be eternal.

About the Author

Alexander A. Bolonkin was born in former USSR. He holds a doctoral degree in Aviation Engineering from Moscow Aviation Institute and a postdoctoral degree in Aerospace Engineering from Leningrad Polytechnic University. He has held the positions senior engineer at the Antonov Aircraft Design Company and chairman of the Reliability Department at the Glushko Rocket Design Company. He has also lectured at the Moscow Aviation Universities. Following his arrival in the United States of America in 1988, he lectured at the New Jersey Institute of Technology and worked as a senior researcher at NASA and the US Air Force Research Laboratories.

Prof. Bolonkin is the author of more than 170 scientific articles and books, and 17 inventions to his credit. His most notable books include: *The Development of Soviet Rocket Engine* (Delphic Ass., Inc., Washington, 1991); *Non-Rocket Space Launch and Flight* (Elsevier, Amsterdam, 2006); *New Concepts, Ideas, Innovation in Aerospace, Technology and Human Life* (NOVA Science Publisher, Inc., New York, 2008); *Macro-Projects: Environment and Technology* (NOVA Science Publisher, Inc., New York, 2009); and *New Technologies and Revolutionary Projects* (Scribd, 2010, 324 pp.).

Part I

Universe. Who Are We?
Where Are We?

1 Macro World

1.1 Universe

The **universe** is commonly defined as the totality of everything that exists, including all physical matter and energy, the planets, stars, galaxies, and the contents of intergalactic space. The term *universe* may be used in slightly different contextual senses, denoting such concepts as the *cosmos*, the *world*, or *nature*.

Observations of earlier stages in the development of the universe, which can be seen at great distances, suggest that the universe has been governed by the same physical laws. The solar system is embedded in a galaxy composed of billions of stars, the Milky Way, and other galaxies exist outside it, as far as astronomical instruments can reach. Careful studies of the distribution of these galaxies and their spectral lines have led the modern cosmology. Discovery of the red shift and cosmic microwave background (CMB) radiation revealed that the universe is expanding and apparently had a beginning. This high-resolution image of the Hubble ultradeep field shows a diverse range of galaxies, each consisting of billions of stars.

According to the prevailing scientific model of the universe, known as the Big Bang, the universe expanded from an extremely hot, dense phase called the Planck epoch, in which all the matter and energy of the observable universe was concentrated. Since the Planck epoch, the universe has been expanding to its present form, possibly with a brief period ($<10^{-32}$ s, 15 billions of years) of cosmic inflation. Several independent experimental measurements support this theoretical expansion and, more generally, the Big Bang theory. Recent observations indicate that this expansion is accelerating because of dark energy, and that most of the matter in the universe may be in a form that cannot be detected by present instruments, and so is not accounted for in the present models of the universe; this has been termed dark matter. The imprecision of current observations has hindered predictions of the ultimate fate of the universe.

Current interpretations of astronomical observations indicate that the age of the universe is 13.75 ± 0.17 billion years, and that the diameter of the observable universe is at least 93 billion light-years, or 8.80×10^{26} m. According to general relativity, space can expand faster than the speed of light, although we can view only a small portion of the universe due to the limitation imposed by light speed. Because we cannot observe space beyond the limitations of light (or any electromagnetic radiation), it is uncertain whether the size of the universe is finite or infinite.

More customarily, the universe is defined as everything that exists, has existed, and will exist. According to this definition and our present understanding, the universe consists of three elements: space and time, collectively known as space-time

Universe, Human Immortality and Future Human Evaluation. DOI: 10.1016/B978-0-12-415801-6.00001-3

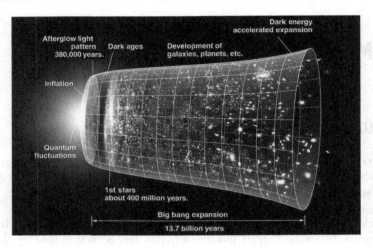

Figure 1.1 Creation and development of the universe.

or the vacuum; matter and various forms of energy and momentum occupying space-time; and the physical laws that govern the first two. A related definition of the term *universe* is everything that exists at a single moment of cosmological time, such as the present, as in the sentence "The universe is now bathed uniformly in microwave radiation."

The universe is very large and possibly infinite in volume. The region visible from Earth (the observable universe) is about 92 billion light-years across, based on where the expansion of space has taken the most distant objects observed. For comparison, the diameter of a typical galaxy is only 30,000 light-years, and the typical distance between two neighboring galaxies is only 3 million light-years. As an example, our Milky Way Galaxy is roughly 100,000 light-years in diameter, and our nearest sister galaxy, the Andromeda Galaxy, is located roughly 2.5 million light-years away. There are probably more than 100 billion (10^{11}) galaxies in the observable universe. Typical galaxies range from dwarfs with as few as 10 million (10^7) stars up to giants with 1 trillion (10^{12}) stars, all orbiting the galaxy's center of mass. Thus, a very rough estimate from these numbers would suggest there are around 1 sextillion (10^{21}) stars in the observable universe; though a 2003 study by Australian National University astronomers resulted in a figure of 70 sextillion (7×10^{22}) (Figures 1.1 and 1.2).

The universe is believed to be mostly composed of dark energy and dark matter, both of which are poorly understood at present. Less than 5% of the universe is ordinary matter, a relatively small contribution (Figure 1.3).

The observable matter is spread uniformly (*homogeneously*) throughout the universe, when averaged over distances longer than 300 million light-years. However, on smaller length scales, matter is observed to form "clumps," that is, to cluster hierarchically; many atoms are condensed into stars, most stars into galaxies, most

(A) (B)

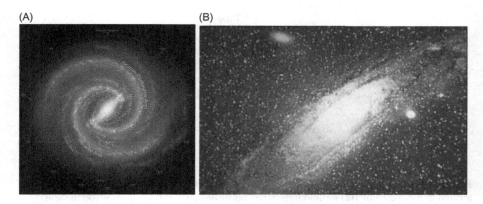

Figure 1.2 (A) Artist's concept of the spiral structure of the Milky Way with two major stellar arms and a bar. (B) Photograph of the "Great Andromeda Nebula" from 1899, later identified as the Andromeda Galaxy.

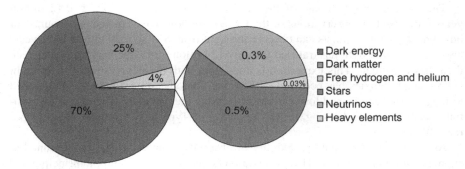

■ Dark energy
□ Dark matter
□ Free hydrogen and helium
■ Stars
□ Neutrinos
□ Heavy elements

Figure 1.3 Composition of the universe.

galaxies into clusters, superclusters, and finally, the largest-scale structures such as the Great Wall of galaxies. The observable matter of the universe is also spread *isotropically*, meaning that no direction of observation seems different from any other; each region of the sky has roughly the same content. The universe is also bathed in a highly isotropic microwave radiation that corresponds to a thermal equilibrium blackbody spectrum of roughly 2.725 K (kelvin). The hypothesis that the large-scale universe is homogeneous and isotropic is known as the cosmological principle, which is supported by astronomical observations.

The present overall density of the universe is very low, roughly 9.9×10^{-30} g/cm^3. This mass-energy appears to consist of 73% dark energy, 23% cold dark matter, and 4% ordinary matter. Thus the density of atoms is on the order of a single hydrogen atom for every 4 m^3 of volume. The properties of dark energy and dark matter are largely unknown. Dark matter gravitates as ordinary matter and thus works to slow the

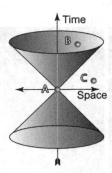

Figure 1.4 Visible and invisible parts of the universe. *B* is visible; *C* is invisible. If a star located at a distance of 100 million light-years burns now, we will see it only 100 million light-years later.

expansion of the universe; by contrast, dark energy accelerates its expansion (Figure 1.3).

The most precise estimate of the universe's age is about 13.73 ± 0.12 billion years old, based on observations of the CMB radiation. This expansion accounts for how Earth-bound scientists can observe the light from a galaxy 30 billion light-years away, even if that light has traveled for only 13 billion years; the very space between them has expanded. This expansion is consistent with the observation that the light from distant galaxies has been red-shifted; the photons emitted have been stretched to longer wavelengths and lower frequency during their journey. The rate of this spatial expansion is accelerating, based on studies of Type Ia supernovae and corroborated by other data.

As such, the conditional probability of observing a universe that is fine-tuned to support intelligent life is 1. This observation is known as the anthropic principle and is particularly relevant if the creation of the universe was probabilistic or if multiple universes with a variety of properties exist (Figure 1.4).

Of the four fundamental interactions, gravitation is dominant at cosmological length scales; that is, the other three forces are believed to play a negligible role in determining structures at the level of planets, stars, galaxies, and larger-scale structures. Given gravitation's predominance in shaping cosmological structures, accurate predictions of the universe's past and future require an accurate theory of gravitation. The best theory available is Albert Einstein's general theory of relativity.

That leads to a single form for the metric tensor, called the Friedmann–Lemaître–Robertson–Walker metric. This metric has only two undetermined parameters: an overall length scale R that can vary with time, and a curvature index k that can be only 0, 1, or −1, corresponding to flat Euclidean geometry, or spaces of positive or negative curvature. In cosmology, solving for the history of the universe is done by calculating R as a function of time, given k and the value of the cosmological constant Λ, which is a (small) parameter in Einstein's field equations. The equation describing how R varies with time is known as the Friedmann equation, after its inventor, Alexander Friedmann (Figure 1.5).

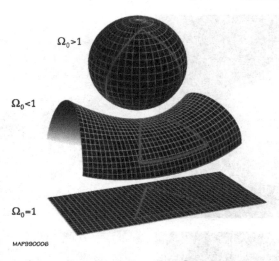

MAP990008

Figure 1.5 Curvature of a two-dimensional plate. Here $\Omega_0 = k$, (52 KB).

The solutions for $R(t)$ depend on k and Λ, but some qualitative features of such solutions are general. First and most importantly, the length scale R of the universe can remain constant *only* if the universe is perfectly isotropic with positive curvature ($k = 1$) and has one precise value of density everywhere, as first noted by Albert Einstein. However, this equilibrium is unstable and because the universe is known to be inhomogeneous on smaller scales, R must change, according to general relativity. Einstein's field equations include a cosmological constant (Λ) that corresponds to an energy density of empty space.

Russian physicist Zel'dovich suggested that Λ is a measure of the zero-point energy associated with virtual particles of quantum field theory, a pervasive vacuum energy that exists everywhere, even in empty space. Evidence for such zero-point energy is observed in the Casimir effect.

The ultimate fate of the universe is still unknown, because it depends critically on the curvature index k and the cosmological constant Λ. If the universe is sufficiently dense, k equals $+1$, meaning that its average curvature throughout is positive and the universe will eventually re-collapse in a Big Crunch, possibly starting a new universe in a Big Bounce. Conversely, if the universe is insufficiently dense, k equals 0 or -1 and the universe will expand forever, cooling off, and eventually becoming inhospitable for all life, as the stars die and all matter coalesces into black holes (the Big Freeze and the heat death of the universe). As noted above, recent data suggest that the expansion speed of the universe is not decreasing as originally expected, but increasing; if this continues indefinitely, the universe will eventually rip itself to shreds (the Big Rip). Experimentally, the universe has an overall density that is very close to the critical value between re-collapse and eternal expansion; more careful astronomical observations are needed to resolve the question (Figure 1.6).

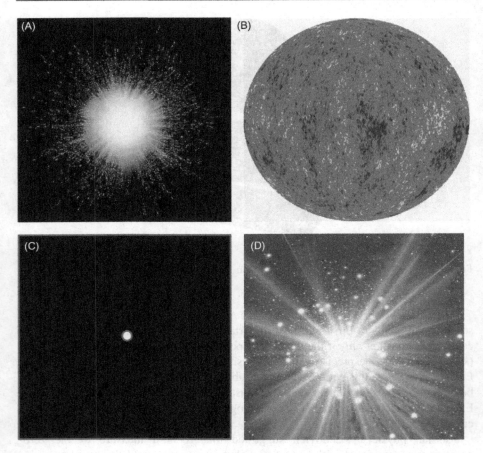

Figure 1.6 History and possible future of the universe: (A) Big Bang, (B) current universe, (C) Big Crunch into dot of singularity, (D) new Big Bang.

1.2 Stars

1.2.1 Conventional Stars

A **star** is a massive, luminous ball of plasma held together by gravity. At the end of its lifetime, a star can also contain a proportion of degenerate matter. The nearest star to Earth is the Sun, which is the source of most of the energy on Earth.

For at least a portion of its life, a star shines due to thermonuclear fusion of hydrogen in its core, releasing energy that traverses the star's interior and then radiates into outer space. Almost all naturally occurring elements heavier than helium were created by stars, either via stellar nucleosynthesis during their lifetimes or via supernova nucleosynthesis when stars explode. Astronomers can determine the mass, age, chemical composition, and many other properties of a star by observing its spectrum, luminosity, and motion through space. The total mass of a star is the principal determinant in its evolution and eventual fate. Other characteristics of a star are determined by its evolutionary history, including diameter, rotation, movement, and temperature. A plot of the temperature of many stars against their luminosities, known as a Hertzsprung–Russell diagram (H–R diagram), allows the age and evolutionary state of a star to be determined.

Binary and multi-star systems consist of two or more stars that are gravitationally bound, and generally move around each other in stable orbits. When two such stars have a relatively close orbit, their gravitational interaction can have a significant impact on their evolution. Stars can form part of a much larger gravitationally bound structure, such as a cluster or a galaxy.

As stars of at least 0.4 solar masses exhaust their supply of hydrogen at their core, their outer layers expand greatly and cool to form a red giant. For example, in about 5 billion years, when the Sun is a red giant, it will expand out to a maximum radius of roughly 1 AU (astronomical unit) (150,000,000 km), 250 times its present size. As a giant, the Sun will lose roughly 30% of its current mass.

An evolved, average-sized star will now shed its outer layers as a planetary nebula. If what remains after the outer atmosphere has been shed is <1.4 solar masses, it shrinks to a relatively tiny object (about the size of Earth) that is not massive enough for further compression to take place, known as a white dwarf. The electron-degenerate matter inside a white dwarf is no longer plasma, even though stars are generally referred to as being spheres of plasma. White dwarfs will eventually fade into black dwarfs over a very long period of time.

Most of the matter in the star is blown away by the supernova explosion (forming nebulae such as the Crab Nebula) and what remains will be a neutron star (which sometimes manifests itself as a pulsar or X-ray burster) or, in the case of the largest stars (large enough to leave a stellar remnant greater than roughly 4 solar masses), a black hole. In a neutron star, the matter is in a state known as neutron-degenerate matter, with a more exotic form of degenerate matter, quantum chromodynamics (QCD) matter, possibly present in the core. Within a black hole, the matter is in a state that is not currently understood.

The nearest star to the Earth, apart from the Sun, is Proxima Centauri, which is 39.9 trillion (10^{12}) km, or 4.2 light-years away. Light from Proxima Centauri takes 4.2 years to reach Earth. Traveling at the orbital speed of the Space Shuttle (5 miles/s—almost 30,000 km/h), it would take about 150,000 years to get there.

Most stars are between 1 billion and 10 billion years old. Some stars may even be close to 13.7 billion years old—the observed age of the universe. The oldest star yet discovered, HE 1523-0901, is an estimated 13.2 billion years old.

The more massive the star, the shorter its lifespan, primarily because massive stars have greater pressure on their cores, causing them to burn hydrogen more rapidly. The most massive stars last an average of about 1 million years, while stars of minimum mass (red dwarfs) burn their fuel very slowly and last tens to hundreds of billions of years (Figure 1.7).

Stars range in size from neutron stars, which vary anywhere from 20 to 40 km in diameter, to supergiants like Betelgeuse in the Orion constellation, which has

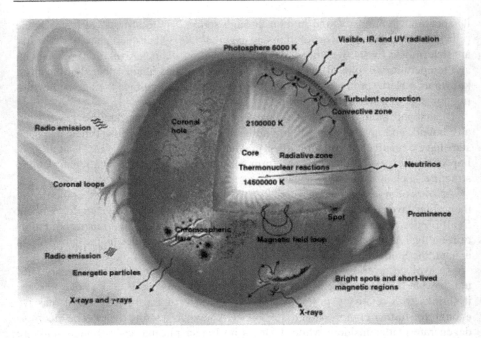

Figure 1.7 The most important star is the Sun.

a diameter approximately 650 times larger than the Sun—about 0.9 billion km. However, Betelgeuse has a much lower density than the Sun.

1.3 Exotic Stars

1.3.1 Neutron Stars

A neutron star is a large gravitationally bound lump of electrically neutral nuclear matter, whose pressure rises from zero (at the surface) to an unknown value in the center.

A neutron star is a type of remnant that can result from the gravitational collapse of a massive star during a Type II, Type Ib, or Type Ic supernova event. Such stars are composed almost entirely of neutrons, which are subatomic particles with zero electrical charge and roughly the same mass as protons. Neutron stars are very hot and are supported against further collapse because of the Pauli exclusion principle. This principle states that no two neutrons (or any other fermionic particle) can occupy the same place and quantum state simultaneously.

A typical neutron star has a mass between 1.35 and about 2.1 solar masses, with a corresponding radius of about 12 km if the Akmal–Pandharipande–Ravenhall (APR) equation of state (EOS) is used. In contrast, the Sun's radius is about 60,000 times that. Neutron stars have overall densities predicted by the APR EOS of 3.7×10^{17}

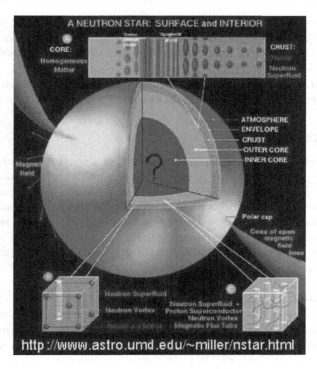

Figure 1.8 Probable structure of neutron star.

$(2.6 \times 10^{14}$ times solar density) to 5.9×10^{17} kg/m³ $(4.1 \times 10^{14}$ times solar density), which compares with the approximate density of an atomic nucleus of 3×10^{17} kg/m³. The neutron star's density varies from below 1×10^{9} kg/m³ in the crust increasing with depth to above 6 or 8×10^{17} kg/m³ deeper inside. This density is approximately equivalent to the mass of the entire human population compressed to the size of a sugar cube.

In general, compact stars of <1.44 solar masses, the Chandrasekhar limit, are white dwarfs; above 2–3 solar masses (the Tolman–Oppenheimer–Volkofflimit), a quark star might be created; however, this is uncertain. Gravitational collapse will always occur on any star over 5 solar masses, inevitably producing a black hole.

The gravitational field at the star's surface is about 2×10^{11} times stronger than that on Earth. The escape velocity is about 100,000 km/s, which is about one-third the speed of light. Such a strong gravitational field acts as a gravitational lens and bends the radiation emitted by the star such that parts of the normally invisible spectrum near the surface become visible (Figure 1.8).

The gravitational binding energy of a 2-solar mass neutron star is equivalent to the total conversion of 1-solar mass to energy (from the law of mass-energy equivalence, $E = mc^2$). That energy was released during the supernova explosion.

A neutron star is so dense that one teaspoon (5 ml) of its material would have a mass over 5×10^{12} kg. The resulting force of gravity is so strong that if an object were to fall from just 1 m high, it would hit the surface of the neutron star at 2000 km/s, or 4.3 million miles/h.

On the basis of current models, the matter at the surface of a neutron star is composed of ordinary atomic nuclei as well as electrons.

1.3.2 White Dwarfs

A **white dwarf**, also called a **degenerate dwarf**, is a small star composed mostly of electron-degenerate matter. It is very dense; a white dwarf's mass is comparable to that of the Sun and its volume is comparable to that of the Earth. Its faint luminosity comes from the emission of stored thermal energy. In January 2009, the Research Consortium on Nearby Stars project counted eight white dwarfs among the hundred nearest star systems of the Sun. The unusual faintness of white dwarfs was first recognized in 1910 by Henry Norris Russell, Edward Charles Pickering, and Williamina Fleming; the name *white dwarf* was coined by Willem Luyten in 1922.

White dwarfs are thought to be the final evolutionary state of all stars whose mass is not high enough to become supernovae—over 97% of the stars in our galaxy. After the hydrogen-fusing lifetime of a main-sequence star of low or medium mass ends, it will expand to a red giant which fuses helium to carbon and oxygen in its core by the triple-alpha process. If a red giant has insufficient mass to generate the core temperatures required to fuse carbon, an inert mass of carbon and oxygen will build up at its center. After shedding its outer layers to form a planetary nebula, it will leave behind this core, which forms the remnant white dwarf. Usually, therefore, white dwarfs are composed of carbon and oxygen. It is also possible that core temperatures suffice to fuse carbon but not neon, in which case an oxygen–neon–magnesium white dwarf may be formed. Also, some helium white dwarfs appear to have been formed by mass loss in binary systems.

The material in a white dwarf no longer undergoes fusion reactions, so the star has no source of energy, nor is it supported by the heat generated by fusion against gravitational collapse. It is supported only by electron degeneracy pressure, causing it to be extremely dense. The physics of degeneracy yields a maximum mass for a nonrotating white dwarf, the Chandrasekhar limit—approximately 1.4 solar masses—beyond which it cannot be supported by degeneracy pressure. A carbon–oxygen white dwarf that approaches this mass limit, typically by mass transfer from a companion star, may explode as a Type Ia supernova via a process known as carbon detonation. (SN 1006 is thought to be a famous example.)

A white dwarf is very hot when it is formed, but as it has no source of energy, it will gradually radiate away its energy and cool down. This means that its radiation, which initially has a high color temperature, will lessen and redden with time. Over a very long time, a white dwarf will cool to temperatures at which it will no longer be visible, and become a cold *black dwarf*. However, because no white dwarf can be older than the age of the universe (approximately 13.7 billion years), even the oldest white dwarfs still radiate at temperatures of a few thousand kelvins, and no black dwarfs are thought to exist yet.

1.3.3 Quark Star

A **quark star** or **strange star** is a hypothetical type of exotic star composed of quark matter, or strange matter. These are ultradense phases of degenerate matter theorized to form inside particularly massive neutron stars.

It is theorized that when the neutron-degenerate matter that makes up a neutron star is put under sufficient pressure due to the star's gravity, the individual neutrons break down into their constituent quarks: up quarks and down quarks. Some of these quarks may then become strange quarks and form strange matter. The star then becomes known as a "quark star" or "strange star," similar to a single gigantic hadron (but bound by gravity rather than the strong force). Quark matter/strange matter is one candidate for the theoretical dark matter that is a feature of several cosmological theories.

1.3.4 Magnetar

A **magnetar** is a type of neutron star with an extremely powerful magnetic field, the decay of which powers the emission of copious amounts of high-energy electromagnetic radiation, particularly X-rays and gamma rays. The theory regarding these objects was proposed by Robert Duncan and Christopher Thompson in 1992, but the first recorded burst of gamma rays thought to have been from a magnetar was detected on March 5, 1979. During the following decade, the magnetar hypothesis became widely accepted as a likely explanation for soft gamma repeaters (SGRs) and anomalous X-ray pulsars (AXPs).

Magnetars are primarily characterized by their extremely powerful magnetic field, which can often reach the order of 10 GT. These magnetic fields are hundreds of thousands of times stronger than any man-made magnet, and quadrillions of times more powerful than the field surrounding Earth. As of 2010, they are the most magnetic objects ever detected in the universe.

A magnetic field of 10 GT is enormous relative to magnetic fields typically encountered on Earth. Earth has a geomagnetic field of 30–60 μT, and a neodymium-based rare earth magnet has a field of about 1 T, with a magnetic energy density of 4.0×10^5 J/m^3. A 10 GT field, by contrast, has an energy density of 4.0×10^{25} J/m^3, with an E/c^2 mass density $>10^4$ times that of lead. The magnetic field of a magnetar would be lethal even at a distance of 1,000 km, tearing tissues due to the diamagnetism of water. At a distance halfway to the moon, a magnetar could strip information from all credit cards on Earth.

As described in the February 2003 *Scientific American*, remarkable things happen within a magnetic field of magnetar strength: "X-ray photons readily split into two or merge together. The vacuum itself is polarized, becoming strongly birefringent, like a calcite crystal. Atoms are deformed into long cylinders thinner than the quantum-relativistic wavelength of an electron." In a field of about 10^5 T atomic orbitals deform into rod shapes. At 10^{10}T, a hydrogen atom becomes a spindle 200 times narrower than its normal diameter.

Although most common magnetic phenomena are electromagnetic, a second source of magnetism occurs due to the spin magnetic moment of subatomic particles. Spin magnetic moment is responsible for the magnetic field of magnetars, and is also exploited in technologies such as nuclear magnetic resonance and magnetic resonance imaging (NMR/MRI).

1.3.5 Black Hole

In general relativity, a black hole is a region of space in which the gravitational field is so powerful that nothing, including light, can escape its pull. The black hole has a one-way surface, called the event horizon, into which objects can fall, but out of which nothing can come. It is called "black" because it absorbs all the light that hits it, reflecting nothing, just like a perfect black body in thermodynamics.

Despite its invisible interior, a black hole can reveal its presence through interaction with other matter. A black hole can be inferred by tracking the movement of a group of stars that orbit a region in space that looks empty. Alternatively, one can see gas falling into a relatively small black hole, from a companion star. This gas spirals inward, heating up to very high temperature and emitting large amounts of radiation that can be detected from earth-bound and earth-orbiting telescopes. Such observations have resulted in the general scientific consensus that, barring a breakdown in our understanding of nature, black holes do exist in our universe.

It is impossible to directly observe a black hole. However, it is possible to infer its presence by its gravitational action on the surrounding environment, particularly with microquasars and active galactic nuclei, where material falling into a nearby black hole is significantly heated and emits a large amount of X-ray radiation. This observation method allows astronomers to detect their existence. The only objects that agree with these observations and are consistent within the framework of general relativity are black holes.

A black hole has only three independent physical properties: mass, charge, and angular momentum.

In astronomy, black holes are classified as:

- *Supermassive*: These contain hundreds of thousands to billions of solar masses and are thought to exist in the center of most galaxies, including the Milky Way.
- *Intermediate*: These contain thousands of solar masses.
- *Micro (also mini black holes)*: These have masses much less than that of a star. At these sizes, quantum mechanics is expected to take effect. There is no known mechanism for them to form via normal processes of stellar evolution, but certain inflationary scenarios predict their production during the early stages of the evolution of the universe.

According to some theories of quantum gravity, black holes may also be produced in the highly energetic reaction produced by cosmic rays hitting the atmosphere or even in particle accelerators such as the Large Hadron Collider. The theory of Hawking radiation predicts that such black holes will evaporate in bright flashes

(A)   (B)

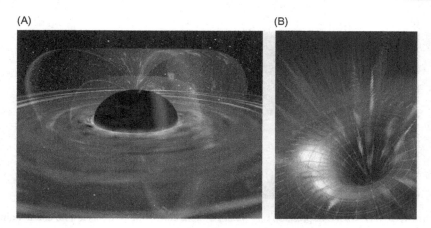

Figure 1.9 (A) Artist's concept of a stellar mass black hole. (B) Artist's rendering showing the space-time contours around a black hole.
Credit: NASA.

of gamma radiation. NASA's Fermi Gamma-ray Space Telescope satellite (formerly GLAST) launched in 2008 is searching for such flashes (Figure 1.9).

The defining feature of a black hole is the appearance of an *event horizon*, a boundary in space-time beyond which events cannot affect an outside observer.

Because the event horizon is not a material surface but rather merely a mathematically defined demarcation boundary, nothing prevents matter or radiation from entering a black hole, only from exiting one.

For a nonrotating (static) black hole, the *Schwarzschild radius* delimits a spherical event horizon. The Schwarzschild radius of an object is proportional to the mass. Rotating black holes have distorted nonspherical event horizons. The description of black holes given by general relativity is known to be an approximation, and it is expected that quantum gravity effects become significant near the vicinity of the event horizon. This allows observations of matter in the vicinity of a black hole's event horizon to be used to indirectly study general relativity and proposed extensions to it (Figure 1.10).

Though black holes themselves may not radiate energy, electromagnetic radiation and matter particles may be radiated from just outside the event horizon via *Hawking radiation*.

At the center of a black hole lies the *singularity*, where matter is crushed to infinite density, the pull of gravity is infinitely strong, and space-time has infinite curvature. This means that a black hole's mass becomes entirely compressed into a region with zero volume. This zero-volume, infinitely dense region at the center of a black hole is called a *gravitational singularity*.

The singularity of a nonrotating black hole has zero length, width, and height; a rotating black hole is smeared out to form a ring shape lying in the plane of rotation. The ring still has no thickness and hence no volume.

The *photon sphere* is a spherical boundary of zero thickness such that photons moving along tangents to the sphere will be trapped in a circular orbit. For nonrotating

(A)   (B)

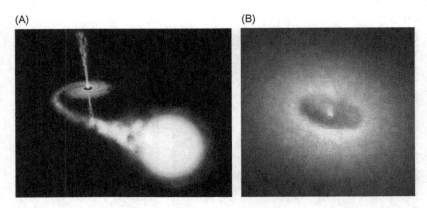

Figure 1.10 (A) Artist's impression of a binary system consisting of a black hole and a main-sequence star. The black hole is drawing matter from the main-sequence star via an accretion disk around it, and some of this matter forms a gas jet. (B) Ring around a suspected black hole in galaxy NGC 4261, November 1992.
Courtesy of Space Telescope Science.

black holes, the photon sphere has a radius 1.5 times the Schwarzschild radius. The orbits are dynamically unstable, hence any small perturbation (such as a particle of infalling matter) will grow over time, setting it either on an outward trajectory, escaping the black hole, or on an inward spiral, eventually crossing the event horizon.

Rotating black holes are surrounded by a region of space-time in which it is impossible to stand still, called the *ergosphere*. Objects and radiation (including light) can stay in orbit within the ergosphere without falling to the center.

Once a black hole has formed, it can continue to grow by absorbing additional matter. Any black hole will continually absorb interstellar dust from its direct surroundings and omnipresent cosmic background radiation.

Much larger contributions can be obtained when a black hole merges with other stars or compact objects.

1.4 Sun

The **Sun** is the star at the center of the solar system. It has a diameter of about 1,392,000 km, about 109 times that of Earth, and its mass (about 2×10^{30} kg, 330,000 times that of Earth) accounts for about 99.86% of the total mass of the solar system. About three-quarters of the Sun's mass consists of hydrogen, while the rest is mostly helium. Less than 2% consists of heavier elements, including oxygen, carbon, neon, iron, and others.

Sunlight is Earth's primary source of energy. The solar constant is the amount of power that the Sun deposits per unit area that is directly exposed to sunlight. The solar constant is equal to approximately 1368 W/m^2 (watts per square meter) at a distance of 1 AU from the Sun (i.e., on or near Earth). Sunlight on the surface of Earth

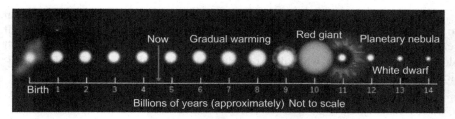

Figure 1.11 Life cycle of the Sun. Evolution of the Sun in time.

is attenuated by the Earth's atmosphere so that less power arrives at the surface—closer to 1000W/m^2 in clear conditions when the Sun is near the zenith.

Solar energy can be harnessed by a variety of natural and synthetic processes—photosynthesis by plants captures the energy of sunlight and converts it to chemical form (oxygen and reduced carbon compounds), while direct heating or electrical conversion by solar cells are used by solar power equipment to generate electricity or to do other useful work, sometimes employing concentrating solar power (that is measured in Suns). The energy stored in petroleum and other fossil fuels was originally converted from sunlight by photosynthesis in the distant past.

The Sun's surface temperature is approximately 5778 K (5505°C). Of the 50 nearest stellar systems within 17 light-years from Earth (the closest being a red dwarf named Proxima Centauri at approximately 4.2 light-years away).

The Sun's resultant velocity with respect to the CMB is about 370 km/s in the direction of Crater or Leo.

The mean distance of the Sun from the Earth is approximately 149.6 million km (1 AU). At this average distance, light travels from the Sun to Earth in about 8 min and 19 s. The energy of this sunlight supports almost all life on Earth by photosynthesis, and drives Earth's climate and weather.

The core of the Sun is considered to extend from the center to about 0.2–0.25 of the solar radius. It has a density of up to 150g/cm^3 (about 150 times the density of water) and a temperature of close to 13.6 million K.

The Earth's fate is precarious. As a red giant, the Sun will have a maximum radius beyond the Earth's current orbit, 1 AU (1.5×10^{11} m), 250 times the present radius of the Sun. However, by the time it is an asymptotic giant branch star, the Sun will have lost roughly 30% of its present mass due to a stellar wind, so the orbits of the planets will move outward. If it were only for this, Earth would probably be spared, but new research suggests that Earth will be swallowed by the Sun owing to tidal interactions. Even if Earth would escape incineration in the Sun, still all its water will be boiled away and most of its atmosphere would escape into space. Even during its current life in the main sequence, the Sun is gradually becoming more luminous (about 10% every 1 billion years), and its surface temperature is slowly rising. The Sun used to be fainter in the past, which is possibly the reason life on Earth has only existed for about 1 billion years on land. The increase in solar temperatures is such that in about another billion years, the surface of the Earth will likely become too hot for liquid water to exist, ending all terrestrial life (Figure 1.11).

Following the red-giant phase, intense thermal pulsations will cause the Sun to throw off its outer layers, forming a planetary nebula. The only object that will remain after the outer layers are ejected is the extremely hot stellar core, which will slowly cool and fade as a white dwarf over many billions of years. This stellar evolution scenario is typical of low- to medium-mass stars.

1.5 Solar Planets

The inner solar system is the traditional name for the region comprising the terrestrial planets and asteroids. The inner planets are Mercury, Venus, Earth, and Mars.

The four inner or terrestrial planets have dense, rocky compositions, few or no moons, and no ring systems. They are composed largely of refractory minerals, such as the silicates which form their crusts and mantles, and metals such as iron and nickel, which form their cores. Three of the four inner planets (Venus, Earth, and Mars) have atmospheres substantial enough to generate weather; all have impact craters and tectonic surface features such as rift valleys and volcanoes. The term *inner planet* should not be confused with *inferior planet*, which designates those planets which are closer to the Sun than the Earth is (i.e., Mercury and Venus).

Venus has no natural satellites. It is the hottest planet, with surface temperatures over 400°C, most likely due to the amount of greenhouse gases in the atmosphere.

Mars (1.5 AU from the Sun) is smaller than Earth and Venus (0.107 earth masses). It possesses an atmosphere of mostly carbon dioxide with a surface pressure of 6.1 mbar (roughly 0.6% that of the Earth's). Its surface, peppered with vast volcanoes such as Olympus Mons and rift valleys such as Valles Marineris, shows geological activity that may have persisted until as recently as 2 million years ago. Its red color comes from iron oxide (rust) in its soil. Mars has two tiny natural satellites (Deimos and Phobos) thought to be captured asteroids.

The four outer planets, or gas giants (sometimes called Jovian planets), collectively make up 99% of the mass known to orbit the Sun. Jupiter and Saturn are each many tens of times the mass of the Earth and consist overwhelmingly of hydrogen and helium; Uranus and Neptune are far less massive (<20 earth masses) and possess more ice in their makeup. For these reasons, some astronomers suggest they belong in their own category, "ice giants." All four gas giants have rings, although only Saturn's ring system is easily observed from Earth. The term *outer planet* should not be confused with *superior planet*, which designates planets outside Earth's orbit and thus includes both the outer planets and Mars.

1.5.1 Jupiter

Jupiter (5.2 AU), at 318 earth masses, is 2.5 times the mass of all the other planets put together. It is composed largely of hydrogen and helium. Jupiter's strong internal heat creates a number of semipermanent features in its atmosphere, such as cloud bands and the Great Red Spot.

Jupiter has 63 known satellites. The four largest, Ganymede, Callisto, Io, and Europa, show similarities to the terrestrial planets, such as volcanism and internal heating. Ganymede, the largest satellite in the solar system, is larger than Mercury.

1.5.2 Saturn

Saturn (9.5 AU), distinguished by its extensive ring system, has several similarities to Jupiter, such as its atmospheric composition and magnetosphere. Although Saturn has 60% of Jupiter's volume, it is less than one-third as massive, at 95 earth masses, making it the least dense planet in the solar system. The rings of Saturn are made up of small ice and rock particles.

Saturn has 62 confirmed satellites; two of which, Titan and Enceladus, show signs of geological activity, though they are largely made up of ice. Titan, the second largest moon in the solar system, is larger than Mercury and the only satellite in the solar system with a substantial atmosphere.

1.5.3 Uranus

Uranus (19.6 AU), at 14 earth masses, is the lightest of the outer planets. Uniquely among the planets, it orbits the Sun on its side; its axial tilt is over 90° to the ecliptic. It has a much colder core than the other gas giants and radiates very little heat into space.

Uranus has 27 known satellites, the largest ones being Titania, Oberon, Umbriel, Ariel, and Miranda.

1.5.4 Neptune

Neptune (30 AU), though slightly smaller than Uranus, is more massive (equivalent to 17 Earths) and therefore more dense. It radiates more internal heat, but not as much as Jupiter or Saturn.

Neptune has 13 known satellites. The largest, Triton, is geologically active, with geysers of liquid nitrogen. Triton is the only large satellite with a retrograde orbit. Neptune is accompanied in its orbit by a number of minor planets, termed Neptune Trojans, that are in 1:1 resonance with it.

1.5.5 Neighborhood

The immediate galactic neighborhood of the solar system is known as the Local Interstellar Cloud or Local Fluff, an area of denser cloud in an otherwise sparse region known as the Local Bubble, an hourglass-shaped cavity in the interstellar medium roughly 300 light-years across. The bubble is suffused with high-temperature plasma that suggests it is the product of several recent supernovae.

There are relatively few stars within 10 light-years (95 trillion km) of the Sun. The closest is the triple-star system Alpha Centauri, which is about 4.4 light-years away. Alpha Centauri A and B are a closely tied pair of sunlike stars, while the small red dwarf Alpha Centauri C (also known as Proxima Centauri) orbits the pair at a distance of 0.2 light-years. The stars next closest to the Sun are the red dwarfs Barnard's Star (at 5.9 light-years), Wolf 359 (7.8 light-years), and Lalande 21185 (8.3 light-years). The largest star within 10 light-years is Sirius, a bright main-sequence star roughly twice the Sun's mass and orbited by a white dwarf called Sirius B. It lies 8.6 light-years away. The remaining systems within 10 light-years are the binary red dwarf system Luyten 726-8 (8.7 light-years) and the solitary red dwarf Ross 154 (9.7 light-years). Our closest solitary sunlike star is Tau Ceti, which lies 11.9 light-years away. It has roughly 80% of the Sun's mass, but only 60% of its luminosity. The closest known extrasolar planet to the Sun lies around the star Epsilon Eridani, a star slightly dimmer and redder than the Sun, which lies 10.5 light-years away. Its one confirmed planet, Epsilon Eridani b, is roughly 1.5 times Jupiter's mass and orbits its star every 6.9 years.

1.6 Summary

1. The universe is commonly defined as the totality of everything that exists. The radius of the universe is about 8.8×10^{26} m, the mass is 1.7×10^{53} kg, and its energy is 1.5×10^{70} J. The universe contains millions (billions) of galaxies. Every galaxy has millions (billions) of stars. Many stars are similar to our Sun and have planets with conditions close to our Earths.
2. The universe is about 13.7 billion years old. That means that intelligent life can be in many (millions) planets of the universe. The population of Earth is about 6.9 billion in 2010. You can estimate what part of the universe are you? What part of the universe life is your life?
3. Intelligent civilizations (whether desired or not) take part in a competition to determine a Supreme Mind. Whoever first reaches the highest knowledge will win this competition and will be the Lord of the Universe (see Part II). It is possible that this highest civilization will create new universes.

2 Micro World

2.1 Matter

Matter is a general term for the substance of which all physical objects are made. Typically, matter includes atoms and other particles that have mass. A common way of defining *matter* is as anything that has mass and occupies volume. In practice, however, there is no single correct scientific meaning of "matter," as different fields use the term in different and sometimes incompatible ways.

A definition of "matter" more fine-scale than the atoms and molecules definition is: *matter is made up of what atoms and molecules are made of*, meaning anything made of protons, neutrons, and electrons.

Matter is commonly said to exist in four *states* (or *phases*): solid, liquid, gas, and plasma. However, advances in experimental techniques have realized other phases, previously only theoretical constructs, such as Bose–Einstein condensates and fermionic condensates. A focus on an elementary-particle view of matter also leads to new phases of matter, such as the quark–gluon plasma.

In physics and chemistry, matter exhibits both wave-like and particle-like properties, the so-called wave–particle duality.

In the realm of cosmology, extensions of the term *matter* are invoked to include dark matter and dark energy, concepts introduced to explain some odd phenomena of the observable universe, such as the galactic rotation curve. These exotic forms of "matter" do not refer to matter as "building blocks," but rather to currently poorly understood forms of mass and energy.

2.2 Atoms, Molecules

Atoms are the smallest (size is about some 10^{-8} m) neutral particles into which matter can be divided by chemical reactions (Figure 2.1). An atom consists of a small, heavy nucleus surrounded by a relatively large, light cloud of electrons. Each type of atom corresponds to a specific chemical element. To date, 117 elements have been discovered (atomic numbers 1–116 and 118), and the first 111 have received official names. The well-known periodic table provides an overview. Atoms consist of protons and neutrons within the nucleus. Within these particles, there are smaller particles which are then made up of even smaller particles.

Molecules are the smallest particles into which a nonelemental substance can be divided while maintaining the physical properties of the substance. Each type of

Universe, Human Immortality and Future Human Evaluation. DOI: 10.1016/B978-0-12-415801-6.00002-5

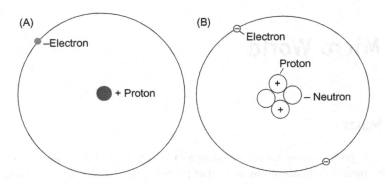

Figure 2.1 (A) A hydrogen atom contains one proton and one electron. (B) A helium atom contains two protons, two neutrons, and two electrons.

molecule corresponds to a specific chemical compound. Molecules are a composite of two or more atoms (Figure 2.3).

Atoms contain small (size is about some 10^{-15} m) nuclei and electrons orbit around these nuclei. The nuclei of most atoms consist of protons and neutrons, which are therefore collectively referred to as nucleons. The number of protons in a nucleus is the atomic number and defines the type of element the atom forms. The number of neutrons determines the isotope of an element. For example, the carbon-12 isotope has 6 protons and 6 neutrons, while the carbon-14 isotope has 6 protons and 8 neutrons.

While bound neutrons in stable nuclei are stable, free neutrons are unstable; they undergo beta decay with a lifetime of just under 15 min. Free neutrons are produced in nuclear fission and fusion. Dedicated neutron sources like research reactors and spallation sources produce free neutrons for use in irradiation and in neutron-scattering experiments.

Outside the nucleus, free neutrons are unstable and have a mean lifetime of 885.7 ± 0.8 s, decaying by emission of a negative electron and antineutrino to become a proton.

This decay mode, known as beta decay, can also transform the character of neutrons within unstable nuclei.

Bound inside a nucleus, protons can also transform via inverse beta decay into neutrons. In this case, the transformation occurs by emission of a positron (antielectron) and a neutrino (instead of an antineutrino).

The transformation of a proton to a neutron inside of a nucleus is also possible through electron capture.

There are four forces active between particles: strong interaction, weak interaction, charge force (Coulomb force), and gravitation force. The strong interaction is the strongest force in short nuclei distance, and the gravitation force is very small into atom.

Beta decay and electron capture are types of radioactive decay and are both governed by the weak interaction (Figures 2.2 and 2.3).

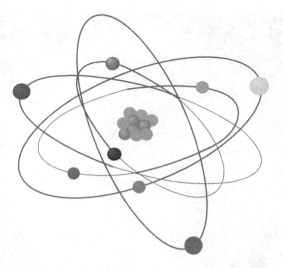

Figure 2.2 A more complex atom that contains many protons, neutrons, and electrons.

Water molecule

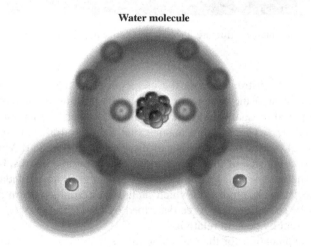

Figure 2.3 A molecule contains some atoms connected by its electrons.

2.3 Nucleus, Protons, Neutrons, Quarks, and Nuclear Forces

The nuclear force is only felt among hadrons. In particle physics, a hadron is a bound state of quarks (particles into nucleus). Hadrons are held together by the strong force, similar to how atoms are held together by electromagnetic force. There are two subsets of hadrons: baryons and mesons; the most well-known baryons are protons and neutrons.

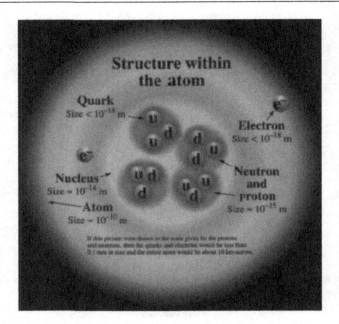

Figure 2.4 Atom and nucleus structure. Protons and neutrons contain quarks.

The Universe is constructed from the elementary particles. Six leptons and six quarks comprise most of the matter; for example, the protons and neutrons of atomic nuclei are composed of quarks, and the ubiquitous electron is a lepton (Figure 2.4). The universe appears to behave in a manner that regularly follows a set of physical laws and physical constants. According to the prevailing Standard Model of physics, all matter is composed of three generations of leptons and quarks, both of which are fermions. These elementary particles interact via at most three fundamental interactions: the electroweak interaction which includes electromagnetism and the weak nuclear force; the strong nuclear force described by QCD; and gravity, which is best described at present by general relativity.

There is no explanation for the particular values that physical constants appear to have throughout our universe, such as Planck's constant h or the gravitational constant G. Several conservation laws have been identified, such as the conservation of charge, momentum, angular momentum, and energy; in many cases, these conservation laws can be related to symmetries or mathematical identities.

At much smaller separations between nucleons, the force is very powerfully repulsive, which keeps the nucleons at a certain average separation. Beyond about 1.7 femtometer (fm) separation, the force drops to negligibly small values.

At short distances, the nuclear force is stronger than the Coulomb force; it can overcome the Coulomb repulsion of protons inside the nucleus. However, the Coulomb force between protons has a much larger range and becomes the only significant force between protons when their separation exceeds about 2.5 fm.

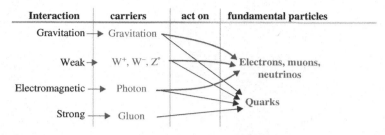

Figure 2.5 Interaction between fundamental particles.

The nuclear force is nearly independent of whether the nucleons are neutrons or protons. This property is called *charge independence*. It depends on whether the spins of the nucleons are parallel or antiparallel, and has a noncentral or *tensor* component. This part of the force does not conserve orbital angular momentum, which is a constant of motion under central forces.

The **nuclear force** (or **nucleon–nucleon interaction** or **residual strong force**) is the force between two or more nucleons. It is responsible for binding of protons and neutrons into atomic nuclei. To a large extent, this force can be understood in terms of the exchange of virtual light mesons, such as the pions. Sometimes the nuclear force is called the residual strong force, in contrast to the strong interactions which are now understood to arise from QCD. This phrasing arose during the 1970s when QCD was being established. Before that time, the *strong nuclear force* referred to the inter-nucleon potential. After the verification of the quark model, *strong interaction* has come to mean QCD.

A **subatomic particle** is an elementary or composite particle smaller than an atom. Particle physics and nuclear physics are concerned with the study of these particles, their interactions, and nonatomic matter.

Elementary particles are particles with no measurable internal structure; that is, they are not composed of other particles. They are the fundamental objects of quantum field theory. Many families and subfamilies of elementary particles exist. Elementary particles are classified according to their spin. Fermions have half-integer spin while bosons have integer spin. All the particles of the Standard Model have been observed, with the exception of the Higgs boson (Figures 2.5 and 2.6).

Subatomic particles include the atomic constituents: electrons, protons, and neutrons. Protons and neutrons are composite particles, consisting of quarks. A proton contains two up quarks and one down quark, while a neutron consists of one up quark and two down quarks; the quarks are held together in the nucleus by gluons. There are six different types of quark in all ("up," "down," "bottom," "top," "strange," and "charm"), as well as other particles, including photons and neutrinos, which are produced copiously in the sun. Most of the particles that have been discovered are encountered in cosmic rays interacting with matter and are produced by scattering processes in particle accelerators. There are dozens of known subatomic particles.

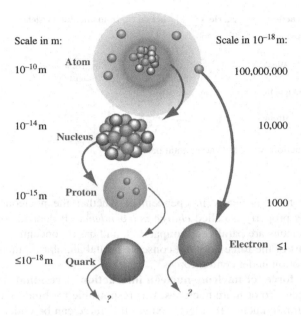

Figure 2.6 Size and scale of nucleus particles.

2.4 Degenerate Matter

Degenerate matter is matter that has such very high density that the dominant contribution to its pressure rises from the Pauli exclusion principle. The pressure maintained by a body of degenerate matter is called the degeneracy pressure, and arises because the Pauli principle forbids the constituent particles to occupy identical quantum states. Any attempt to force them close enough together that they are not clearly separated by position must place them in different energy levels. Therefore, reducing the volume requires forcing many of the particles into higher-energy quantum states. This requires additional compression force and is manifest as a resisting pressure.

Degenerate matter still has normal thermal pressure, but at high densities the degeneracy pressure dominates. Thus, increasing the temperature of degenerate matter has a minor effect on total pressure until the temperature rises so high that thermal pressure again dominates total pressure.

Exotic examples of degenerate matter include neutronium, strange matter, metallic hydrogen, and white dwarf matter. Degeneracy pressure contributes to the pressure of conventional solids, but these are not usually considered to be degenerate matter as a significant contribution to their pressure is provided by the interplay between the electrical repulsion of atomic nuclei and the screening of nuclei from each other by electrons allocated among the quantum states determined by the nuclear electrical potentials. In metals, it is useful to treat the conduction electrons alone as a degenerate, free electron gas, while the majority of the electrons are regarded as occupying bound quantum states. This contrasts with the case of the degenerate matter that forms the body of a white dwarf, where all the electrons would be treated as occupying free particle momentum states.

Neutrons are the most "rigid" objects known: their Young's modulus (or more accurately, bulk modulus) is 20 orders of magnitude larger than that of diamond.

For white dwarfs, the degenerate particles are the electrons, while for neutron stars, the degenerate particles are neutrons. In a degenerate gas, when the mass is increased, the pressure is increased, and the particles become spaced closer together, so the object becomes smaller. Degenerate gas can be compressed to very high densities, typical values being in the range of 10^7 g/cm^3.

Preons are subatomic particles proposed to be the constituents of quarks, which become composite particles in preon-based models.

2.5 Summary

1. Matter contains molecules (10^{-9} m). Molecules contain atoms (10^{-10} m). Atoms have a nuclei (10^{-15} m) and electrons. A nucleus is composed of protons and neutrons. Protons and neutrons include quarks (10^{-18} m).
2. The forces that keep the protons and neutrons into nuclei are gigantic but short-acting.
3. The forces that keep together quarks into subatomic particles are very large and increase via distance. At the present time, we cannot separate the quarks into matter particles.

3 Surprising Properties of Our Universe

3.1 All World (Universe) in Point, All Time (Universe History) in Moment

3.1.1 Universe into Black Hole

General notes: Light speed. In 1887, an experiment performed by scientists Albert Michelson and Edward Morley showed that light speed is the fastest speed in the world and nothing can have a speed of more than 300,000 km/s. This fact created a revolution in physics. Based on this experiment, Lorentz showed if the nonaccelerating observer measures the moving object, the length and time in this body will decrease and mass will increase. In 1905, Einstein proposed that the speed of light in a vacuum, measured by a nonaccelerating observer, is independent of the motion of the source or observer. Using this and the principle of relativity as a basis, he derived the special theory of relativity, in which the speed of light in vacuum c featured as a fundamental parameter, also appearing in contexts unrelated to light.

Light and all other electromagnetic radiation always travel at this speed in empty space (vacuum), regardless of the motion of the source or the inertial frame of the observer. Its value is exactly 299,792,458 m/s (approximately 186,282 m/s). In the theory of relativity, c connects space and time, and appears in the famous equation of mass–energy equivalence, $E = mc^2$. The speed of light is the speed of all massless particles and associated fields in a vacuum, and it is predicted by the current theory to be the speed of gravity and of gravitational waves and an upper bound on the speed at which energy, matter, and information can travel.

In most practical cases, light can be thought of as moving instantaneously, but for long distances and very sensitive measurements, the finite speed of light can have noticeable effects. In communicating with distant space probes, it can take minutes to hours for the message to get from Earth to the satellite and back. The light we see from stars left the stars many (millions) years ago, allowing us to study the history of the universe by looking at distant objects. The finite speed of light also limits the theoretical maximum speed of computers, because information must be sent within the computer chips and from chip to chip. Finally, the speed of light can be used with time of flight measurements to measure large distances to high precision.

Albert Einstein postulated that the speed of light in a vacuum was independent of the source or inertial frame of reference, and explored the consequences of that

Universe, Human Immortality and Future Human Evaluation. DOI: 10.1016/B978-0-12-415801-6.00003-7

postulate, deriving the theory of special relativity and showing that the parameter c had relevance outside of the context of light and electromagnetism. After centuries of increasingly precise measurements, in 1975 the speed of light was known to be 299,792,458 m/s with a relative measurement uncertainty of 4 parts per billion. In 1983, the meter was redefined in the International System of Units (SI) as the distance traveled by light in a vacuum in 1/299,792,458 of a second. As a result, the numerical value of c in meters per second is now fixed exactly by the definition of the meter.

Special relativity incorporates the principle that the speed of light is the same for all inertial observers, regardless of the state of motion of the source.

This theory has a wide range of consequences that have been experimentally verified, including counterintuitive ones such as length contraction, time dilation, and relativity of simultaneity, contradicting the classical notion that the duration of the time interval between two events is equal for all observers. (On the other hand, it introduces the space–time interval, which *is* invariant.) Combined with other laws of physics, the two postulates of special relativity predict the equivalence of matter and energy, as expressed in the mass–energy equivalence formula $E = mc^2$, where c is the speed of light in a vacuum. The predictions of special relativity agree well with Newtonian mechanics in their common realm of applicability, specifically in experiments in which all velocities are small compared with the speed of light. Special relativity reveals that c is not just the velocity of a certain phenomenon—namely, the propagation of electromagnetic radiation (light)—but rather a fundamental feature of the way space and time are unified as space–time. One of the consequences of the theory is that it is impossible for any particle that has rest mass to be accelerated to the speed of light.

Lorentz transformation. In physics, the Lorentz transformation, named after the Dutch physicist Hendrik Lorentz, describes how, according to the theory of special relativity, two observers' varying measurements of space and time can be converted into each other's frames of reference. It reflects the surprising fact that observers moving at different velocities may measure different distances, elapsed times, and even different orderings of events.

The Lorentz transformation was originally the result of attempts by Lorentz and others to explain how the speed of light was observed to be independent of the reference frame, and to understand the symmetries of the laws of electromagnetism. Albert Einstein later re-derived the transformation from his postulates of special relativity. The Lorentz transformation supersedes the Galilean transformation of Newtonian physics, which assumes an absolute space and time. According to special relativity, this is a good approximation only at relative speeds much smaller than the speed of light.

The theory of special relativity shows: if we have the two systems (two observers) fixed and moving ($'$), the time, length, and mass measured by observers of the first system (fixed) in the second (moving) system will be different. The values in the moving system ($'$) may be computed by the following equation (Lorentz transformation):

$$l' = l\sqrt{1 - v^2/c^2}, \quad \Delta t' = \sqrt{1 - v^2/c^2}\,\Delta t, \quad m = \frac{m_0}{\sqrt{1 - v^2/c^2}},$$

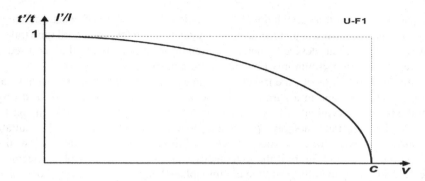

Figure 3.1 Relative time and length of a moving object measured by motionless observer decreases up to zero.

where l is length in m; v is speed in m/s, $v < c$; $c \approx 300,000$ km/s is speed in m/s; Δt is time in s; m is mass in kg.

As the reader can see, the distance in the moving system decreases; the moving clock (in system $(')$) runs more slowly than in the fixed system; the mass in the moving system increases. The last fact is not surprising because the kinetic energy of the moving mass increases. But mass is promotional energy.

Lorentz thought that this was only the mathematical transformation of a system. Einstein believed that was reality. Many people and scientists did not believe in it, because it is out of our logic and Earth experience. However, later experiments showed Einstein was correct (Figure 3.1).

The mass of visible stars in the universe is about $M = 1.7 \times 10^{53}$ kg. That mass is only 0.5% from total mass (dark energy, dark matter, black hole, stars, free hydrogen, neutrino, etc.). The total mass of the universe is about 3.4×10^{55} kg. The radius (event horizon) of a black hole equals $r = 1.5 \times 10^{-27}M = 1.5 \times 10^{-27} \times 3.4 \times 10^{55} = 5.1 \times 10^{28}$ m. This radius is $100 \div 400$ times more than a known radius of our universe $R = (1.2 \div 4.4) \times 10^{26}$ m. That means our universe is located inside the universe black hole! From this follows: Our universe is a closet system. The civilizations of outer universes can come to us, but we cannot go out; we cannot send any information to outer universes.

3.2 Travel into the Future

General notes: Time travel is the concept of moving between different points in time in a manner analogous to moving between different points in space, either sending objects (or in some cases just information) backward in time to some moment before the present, or sending objects forward from the present to the future without the need to experience the intervening period (at least not at the normal rate).

Although time travel has been a common plot device in fiction since the nineteenth century, and one-way travel into the future is arguably possible, given the phenomenon of time dilation based on velocity in the theory of special relativity (exemplified by the twin paradox), as well as gravitational time dilation in the theory of general relativity, it is currently unknown whether the laws of physics would allow backward time travel.

Any technological device, whether fictional or hypothetical, that is used to achieve time travel is commonly known as a time machine.

Some interpretations of time travel also suggest that an attempt to travel backward in time might take one to a parallel universe whose history would begin to diverge from the traveler's original history after the moment the traveler arrived in the past.

Some theories, most notably special and general relativity, suggest that suitable geometries of space–time, or specific types of motion in space, might allow time travel into the past and future if these geometries or motions are possible. In technical papers, physicists generally avoid the commonplace language of "moving" or "traveling" through time ("movement" normally refers only to a change in spatial position as the time coordinate is varied), and instead discuss the possibility of closed time-like curves, which are world-lines that form closed loops in space–time, allowing objects to return to their own past. There are known to be solutions to the equations of general relativity that describe space–times which contain closed time-like curves (such as Gödel space–time), but the physical plausibility of these solutions is uncertain.

Relativity states that if one were to move away from the Earth at relativistic velocities and return, more time would have passed on Earth than for the traveler, so in this sense it is accepted that relativity allows "travel into the future" (according to relativity, there is no single objective answer to how much time has "really" passed between the departure and the return, but there is an objective answer to how much proper time has been experienced by both the Earth and the traveler, i.e., how much each has aged; this is also known as the twin paradox). On the other hand, many in the scientific community believe that backward time travel is highly unlikely. Any theory that would allow time travel would require that problems of causality be resolved. The classic example of a problem involving causality is the "grandfather paradox": what if one were to go back in time and kill one's own grandfather before one's father was conceived? But some scientists believe that paradoxes can be avoided, either by appealing to the Novikov self-consistency principle or to the notion of branching parallel universes.

Stephen Hawking once suggested that the absence of tourists from the future constitutes an argument against the existence of time travel—a variant of the Fermi paradox. Of course, this would not prove that time travel is physically impossible, because it might be that time travel is physically possible but that it is never in fact developed (or is cautiously never used); and even if it is developed, Hawking notes elsewhere that time travel might only be possible in a region of space–time that is warped in the right way, and that if we cannot create such a region until the future, then time travelers would not be able to travel back before that date, so "This picture would explain why we haven't been overrun by tourists from the future." Carl Sagan also once suggested the possibility that time travelers could be here, but are disguising their existence or are not recognized as time travelers.

There are various ways in which a person could "travel into the future" in a limited sense: the person could set things up so that in a small amount of his own subjective time, a large amount of subjective time has passed for other people on Earth. For example, an observer might take a trip away from the Earth and back at relativistic velocities, with the trip only lasting a few years according to the observer's own clocks, and return to find that thousands of years had passed on Earth. It should be noted, though, that according to relativity, there is no objective answer to the question of how much time "really" passed during the trip; it would be equally valid to say that the trip had lasted only a few years or that the trip had lasted thousands of years, depending on your choice of reference frame.

This form of "travel into the future" is theoretically allowed (and has been demonstrated at very small time scales) using the following methods:

- Using velocity-based time dilation under the theory of special relativity, for instance:
  - Traveling at almost the speed of light to a distant star, then slowing down, turning around, and traveling at almost the speed of light back to Earth (the twin paradox).
- Using gravitational time dilation under the theory of general relativity, for instance:
  - Residing inside of a hollow, high-mass object.
  - Residing just outside of the event horizon of a black hole, or sufficiently near an object whose mass or density causes the gravitational time dilation near it to be larger than the time dilation factor on Earth.

Additionally, it might be possible to see the distant future of the Earth using methods which do not involve relativity at all, although it is even more debatable whether these should be deemed a form of "time travel"; these are the concepts of hibernation and suspended animation.

3.3 The Law of Existence of the Universe, the Aim of the Universe, and the Purpose of Humanity

From this follows the possible **aim of universe** (**law of existence of the universe**): *The universe must create the Supreme Brain which will have enough knowledge for the design of the next universe.*

The universe uses intelligent civilizations for this. One makes the competition between them: who creates the first Supreme Mind. The other civilizations become the slaves of forward civilizations as all Earth flora and fauna are used by humanity for our needs.

From here follows the **purpose of humanity**: *Mankind must be the first to create the Supreme Mind.* It is better if one universe can pass the knowledge to a Supreme Mind of the next universe. But as we know now, the eventually recollapse in a Big Crunch, possibly starting a new universe in a Big Bounce (or next Big Bang), will destroy any information and Supreme Mind.

3.4 Summary

From these transformations follow the surprising results:

1. *If the speed of an astronaut aspires (seeks) to the light speed, that the time (millions of years) aspires to a moment and the space (universe) aspires to point.*

In our Earth system, the light from the far stars comes to us over millions of years. In the system of a moving observer, that may be some seconds and all stars will be nearer. Those who had success in school mathematics must remember the definition of an infinity value. That is a value which is more than any given number. But if you divide one (unit) by the infinite you get the zero, or more exactly, the magnitude is close to zero or as close as possible.

2. *Our universe is located inside the universe black hole*! The civilizations of outer universes can come to us, but we cannot go out.

The text above suggests we can move into the future, but it will be the real future. We cannot return in the present time. We will not be useful to the future generation. Our technology will be very old and of poor quality for people of the future. Our information storages are developed well in the present, and future generations will know more about our time than we can save in our brain. Our unilateral relocation may be useful only for us. For example, if a person is sick with an incurable disease, he may die soon. In this case, he would want to relocate into the future because he hopes the future medicine can cure him. Or he wants to know who his children, grandchildren, descendants, and friends will be, or what will be history of his country and the Earth. Or how his ideas, theories, and inventions will be developed and so on. There are persons who freeze themselves after death. They hope the future medicine can bring them back to life and they can continue their life.

But the reader must remember that that will be the real future. You cannot return to your previous life, you cannot change anything. You can only regret your wrong actions. **Causality** is the relationship between an event (the *cause*) and a second event (the *effect*), where the second event is a consequence of the first. There is law of "cause and effect" in physics. The effect cannot take place earlier than the cause.

The reader may ask: how will it be possible to relocate in future?

We must reach a very high speed for it, the speed close to the speed of light, 300,000 km/s. Gigantic energy is needed for it. We need as minimum the generator that converts the matter to energy according the Einstein's equation $E = mc^2$ and a photon rocket (see, for example, the author's scientific work "Converting of Any Matter to Energy by AB-Generator" http://www.scribd.com/doc/24048466/).

Production of this gigantic energy is impossible at the present time. The author offers the following way to reach temporarily the speed close to light speed: spaceship flights to the nearest black hole. The gravity of the black hole accelerates the spaceship to the speed close to 300,000 km/s. If trajectory is designed correctly, the spaceship walks around the black hole in the hyperbolic trajectory and returns to start. The jump to the future depends on the relativistic speed and on the distance to the nearest black hole. The time of jump can be computed.

4 What Is God?

4.1 General Notes: God

God is the English name given to a singular being in theistic and deistic religions (and other belief systems) who is either the *sole* deity in monotheism or a *single* deity in polytheism.

God is most often conceived of as the supernatural creator and overseer of the universe. Theologians have ascribed a variety of attributes to the many different concepts of God. The most common among these include omniscience (infinite knowledge), omnipotence (unlimited power), omnipresence (present everywhere), omnibenevolence (perfect goodness), divine simplicity, and eternal and necessary existence.

God has also been conceived as being incorporeal (immaterial), a personal being, the source of all moral obligation, and the "greatest conceivable existent." These attributes were all supported to varying degrees by the early Jewish, Christian, and Muslim theologian philosophers, including Maimonides, Augustine of Hippo, and Al-Ghazali, respectively. Many notable medieval philosophers and modern philosophers developed arguments for the existence of God. Many notable philosophers and intellectuals have, in contrast, developed arguments *against* the existence of God.

Concepts of God vary widely. Theologians and philosophers have studied countless concepts of God since the dawn of civilization. The Abrahamic concepts of God include the monotheistic definition of God in Judaism, the trinitarian view of Christians, and the Islamic concept of God. The dharmic religions differ in their view of the divine: views of God in Hinduism vary by region, sect, and caste, ranging from monotheistic to polytheistic to atheistic; the view of God in Buddhism is almost nontheist. In modern times, some more abstract concepts have been developed, such as process theology and open theism. Concepts of God held by individual believers vary so widely that there is no clear consensus on the nature of God. The contemporaneous French philosopher Michel Henry has however proposed a phenomenological approach and definition of God as phenomenological essence of life.

4.2 Existence of God

Many arguments that attempt to prove or disprove the existence of God have been proposed by philosophers, theologians, and other thinkers for many centuries. In philosophical terminology, such arguments concern schools of thought on the epistemology of the ontology of God.

Universe, Human Immortality and Future Human Evaluation. DOI: 10.1016/B978-0-12-415801-6.00004-9

Figure 4.1 (A) Image of God. (B) Detail of Sistine Chapel fresco *Creation of the Sun and Moon* by Michelangelo (c. 1512), a well-known example of the depiction of God the Father in Western art.

There are many philosophical issues concerning the existence of God. Some definitions of God are nonspecific, while other definitions can be self-contradictory. Arguments for the existence of God typically include metaphysical, empirical, inductive, and subjective types, while others revolve around perceived holes in evolutionary theory and order and complexity in the world. Arguments against the existence of God typically include empirical, deductive, and inductive types. Conclusions reached include: "God does not exist" (strong atheism); "God almost certainly does not exist" (*de facto* atheism); "no one knows whether God exists" (agnosticism); "God exists, but this cannot be proven or disproven" (weak theism); and "God exists and this can be proven" (strong theism). There are numerous variations on these positions.

Some theologians, such as the scientist and theologian A.E. McGrath, argue that the existence of God cannot be adjudicated for or against by using scientific method. Agnostic Stephen Jay Gould argues that science and religion are not in conflict and do not overlap (nonoverlapping magisteria) (Figure 4.1).

As of 2000, approximately 53% of the world's population identified with one of the three primary Abrahamic religions (33% Christian, 20% Islam, <1% Judaism), 6% with Buddhism, 13% with Hinduism, 6% with traditional Chinese religion, 7% with various other religions, and <15% as nonreligious. Most of these religious beliefs involve a god or gods. (Other Abrahamic religions include Baha'i, Samaritanism, the Rastafari movement, Yazidism, and the Unification Church.)

The author is a scientist. If God exists, any real scientists will study him. They will find the persuasive proof of God's existence, answering the questions: Who is he? Where is he located? What is his age? What is his power? And so on.

But there are no strong scientific arguments about the existence of God. If God is powerful, what is his power?

For example, man is God over ants. Man can destroy an anthill and kill all the ants in a given area. Or he can create excellent conditions for ants to live and flourish. The author thinks if God exists, he is the Supreme Mind of a previous universe, who created our universe and gives us the possibility to create our own Supreme Mind (God) and to continue the existence of high intellect. The life of the universe as the life of any system is limited.

5 What Is the Human Soul?

When the author debates electronic immortality, religious people have trouble regarding the soul. Let us consider this problem.

5.1   General Note: Soul

The soul, in certain spiritual, philosophical, and psychological traditions, is the incorporeal essence of a person or living thing. Many philosophical and spiritual systems teach that humans are souls; some attribute souls to all living things and even to inanimate objects (such as rivers); this belief is commonly called animism. The soul is often believed to exit the body and live on after a person's death, and some religions posit that God creates souls. The soul has often been deemed integral or essential to consciousness and personality, and *soul* sometimes functions as a synonym for *spirit*, *mind*, or *self*, although the soul is said to function in a distinct enough way from both the spirit and the psyche that the terms should not be treated interchangeably.

5.2   Soul in Religions

Christianity. The majority of Christians understands the soul as an ontological reality distinct from, yet integrally connected with, the body. Its characteristics are described in moral, spiritual, and philosophical terms. When people die, their souls will be judged by God and will be determined to spend an eternity in heaven or in hell. Though all branches of Christianity—Catholics, Eastern Orthodox and Oriental Orthodox, evangelical or mainline Protestants—teach that Jesus Christ plays a decisive role in the salvation process, the specifics of that role and the part played by individual persons or ecclesiastical rituals and relationships is a matter of wide diversity in official church teaching, theological speculation, and popular practice. Some Christians believe that if one has not repented of one's sins and trusted in Jesus Christ as Lord and Savior, one will go to hell and suffer eternal separation from God. Variations also exist on this theme, for example, some hold that the unrighteous soul will be destroyed instead of suffering eternally (Annihilationism). Believers will inherit eternal life in heaven and enjoy eternal fellowship with God. There is also a belief that babies (including the unborn) and those with cognitive or mental impairments who have died will be received into heaven on the basis of God's grace through the sacrifice of Jesus.

Universe, Human Immortality and Future Human Evaluation. DOI: 10.1016/B978-0-12-415801-6.00005-0

Buddhism teaches that all things are in a constant state of flux: all is changing, and no permanent state exists by itself. This applies to human beings as much as to anything else in the cosmos. Thus, a human being has no permanent self. According to this doctrine of *anatta* (Pali; Sanskrit: *an tman*—"no-self" or "no soul"), the words "I" or "me" do not refer to any fixed thing. They are simply convenient terms that allow us to refer to an ever-changing entity.

Judaism. The Hebrew terms נפש *nephesh*, חור *ruach* (literally "wind"), and המישנ *neshama* (literally "breath") are used to describe the soul or spirit. The soul is believed to be given by God to a person by his/her first breath, as mentioned in Genesis, "And the LORD God formed man [of] the dust of the ground, and breathed into his nostrils the breath of life; and man became a living soul" (Genesis 2:7). From this statement, the rabbinical interpretation is often that human embryos do not have souls, though abortion is often opposed by the orthodoxy as a form of birth control. Judaism relates the quality of one's soul to one's performance of mitzvot and reaching higher levels of understanding, and thus closeness to God.

In **Hinduism**, the Sanskrit words most closely corresponding to soul are "Jeeva," "Atmaan," and "Purusha," meaning the individual self. The term "soul" is misleading as it implies an object possessed, whereas self signifies the subject which perceives all objects. This self is held to be distinct from the various mental faculties such as desires, thinking, understanding, reasoning, and self-image (ego), all of which are considered to be part of Prakriti (nature).

Islam. Allah narrated in the Quran: "And they ask you (O Muhammad SAW) concerning the Ruh (the Spirit); Say: The Ruh (the Spirit): is the AMAR of your Creator. And of knowledge, you (mankind) have been given only a little" (Noble Quran, Surat Al'Isra', Verse 85) (Figure 5.1).

Further in the Quran, the definition of "Amar" is "And the AMAR of your Creator is such that when it intends for something to happen, it says 'Let it be' then it is done."

Combining the two verses; Ruh (soul, spirit) is that creation of Allah that has the knowledge and authority to use that knowledge.

The definition of that knowledge is also given in the Quran, surah "Al-Baqrah" Chapter 2, verses 31–32, when Allah tells angels about the creation of the first soul, Adam: "That I want to appoint a representative on the Earth ..." (Chapter 2, verse 31) further, Allah says, "And we gave him the knowledge of all the attributes" (Chapter 2, verse 32).

Now this verse explains very clearly about the importance of the soul (Ruh). It has the knowledge of all the attributes that Allah wanted to be revealed; and because we know that a portion of infinity is also an infinity, therefore that knowledge is also boundless. And all this is present in soul, the real human being, without which the physical body is nothing. The physical body is similar to our clothes; when the soul puts it off, it is useless. Hence, it is the soul that is the real you and me, the physical body is just clothing.

Science and medicine seek naturalistic accounts of the observable natural world. This stance is known as methodological naturalism. Much of the scientific study relating to the soul has involved investigating the soul as an object of human belief, or as a concept that shapes cognition and an understanding of the world, rather than as an entity in and of itself.

Figure 5.1 "A soul brought to heaven" by William Bouguereau.

5.3 Summary

1. In the author's opinion, the soul is only the knowledge stored in our brain (see Part II of this book). More than 99% of it is the permanent knowledge about our own history (all that we have seen, heard, spoken, felt, and done), knowledge received in school, college, university, from books, TV, radio, and so on. A very small part (<1%) is variable, our own opinion about people or their actions. That part is changed by new knowledge, information, and situations.

2. Most information is received in our brain from our sensory organs: eyes, ears, tongue, skin, muscles. This information may be easily intercepted by the current sensors, written and stored by current electronic devices. That means our soul may be written and kept safe for a very long time. It may be loaded in new electronic chips (brain). If this electronic brain is provided, the sensors—micro-video camera, microphone, audio speaker, TV screen, and the executive bodies (artificial legs, arms), we get the immortal man (E-being) who continues the life of the real man and will have his initial soul (see Part II).

3. This artificial man will have gigantic advantages over his ancestor. He will not need a body, sleep, or housing. He can live in space and on the bottom of an ocean, travel to other planets by a laser beam with light speed, and be indestructible by any weapon because he will save himself in special storage and restore himself at any time (see Part II).

4. This situation is well described in beautiful fiction book titled *Price of Immortality* by Igor Getmansky (Moscow, ЭКСМО, 2003). This book is only published in Russian. More detailed description of human soul is in the author's article "Science, Soul, Paradise, and Artificial Intelligence" (1999) found in Part II of this book.

6 What Is "I"? What Are "We"?

6.1 Individual (Person)

A **person** (from Latin: *persona*, meaning "mask") is most broadly defined as any individual self-conscious or rational being, or any entity having rights and duties; or often more narrowly defined as an individual human being in particular.

The direct plural of "person" is "persons." The term **people** is the general plural of "person," and is used to refer to several persons plurally in a range from "a few people" up to "all people." "People" is often used to refer to an entire nation or ethnic group, and in this context, "people" can be used as a singular to refer to specific ethnic or national groups (i.e., "a people").

The term **personhood** refers to the state or condition of being an individual person, the essential meaning and constituent properties of what it is to be a person.

While in common parlance "person" and "human" are effectively synonyms, specific fields such as philosophy, law, and others use the term with specialized context-specific meanings.

In philosophy, "person" may apply to any human or nonhuman agent who is regarded as self-conscious and capable of certain kinds of higher-level thought; for example, individuals who have the power to reflect upon and choose their actions.

Boethius gives the definition of "person" as "an individual substance of a rational nature" ("*Naturæ rationalis individua substantia*").

Peter Singer defines "person" simply as being a conscious, thinking being. He also notes that a person must know that they are a person (self-awareness).

Philosopher Thomas I. White argues that the criteria for a person are as follows: (1) is alive; (2) is aware; (3) feels positive and negative sensations; (4) has emotions; (5) has a sense of self; (6) controls its own behavior; (7) recognizes other persons and treats them appropriately; and (8) has a variety of sophisticated cognitive abilities. While many of White's criteria are somewhat anthropocentric, some animals such as dolphins would be considered persons.

Various specific philosophical debates focus on questions about the personhood of different classes of entities.

Some philosophers and those involved in animal welfare, ethology, animal rights, and related subjects consider that certain animals should also be granted personhood. Commonly named species in this context include the great apes, cetaceans, and elephants, because of their apparent intelligence and intricate social rules. The idea of extending personhood to all animals has the support of legal scholars such as Alan Dershowitz and Laurence Tribe of Harvard Law School, and animal law courses are now taught in 92 out of 180 law schools in the United States. On May 9, 2008,

Universe, Human Immortality and Future Human Evaluation. DOI: 10.1016/B978-0-12-415801-6.00006-2

Columbia University Press published "Animals as Persons: Essays on the Abolition of Animal Exploitation" by Prof. Gary L. Francione of Rutgers University School of Law, a collection of writings that summarizes his work to date and makes the case for nonhuman animals as persons.

On the other hand, some proponents of human exceptionalism (also referred to by its critics as speciesism) have countered that we must institute a strict demarcation of personhood based on species membership in order to avoid the horrors of genocide (based on propaganda dehumanizing one or more ethnicities) or the injustices of forced sterilization (as occurred in many countries to people with low IQ scores and prisoners).

Hypothetical beings. Speculatively, there are several other likely categories of beings where personhood might be at issue.

Artificial life. If artificial intelligence (AI), intelligent and self-aware systems of hardware and software are eventually created, what criteria would be used to determine their personhood? Likewise, at what point might human-created biological life be considered to have achieved personhood?

6.2 Intelligence

Intelligence is an umbrella term describing a property of the mind including related abilities, such as the capacities for abstract thought, understanding, communication, reasoning, learning, learning from past experiences, planning, and problem solving.

Intelligence is most widely studied in humans but is also observed in animals and plants.

AI is the intelligence of machines or the simulation of intelligence in machines.

Numerous definitions of and hypotheses about intelligence have been proposed since before the twentieth century, with no consensus yet reached by scholars. Within the discipline of psychology, various approaches to human intelligence have been adopted, with the psychometric approach being especially familiar to the general public. Influenced by his cousin Charles Darwin, Francis Galton was the first scientist to propose a theory of general intelligence; that intelligence is a true, biologically based mental faculty that can be studied by measuring a person's reaction times to cognitive tasks. Galton's research in measuring the head sizes of British scientists and laymen led to the conclusion that head size is unrelated to a person's intelligence.

Alfred Binet and the French school of intelligence believed intelligence was an aggregate of dissimilar abilities, not a unitary entity with specific, identifiable properties.

Animal and plant intelligence. Although humans have been the primary focus of intelligence researchers, scientists have also attempted to investigate animal intelligence, or more broadly, animal cognition. These researchers are interested in studying both mental ability in a particular species and comparing abilities between species. They study various measures of problem solving, as well as mathematical and language abilities. Some challenges in this area are defining intelligence so that it means the same thing across species (e.g., comparing intelligence between literate humans and illiterate animals), and then operationalizing a measure that accurately compares mental ability across different species and contexts.

Wolfgang Köhler's pioneering research on the intelligence of apes is a classic example of research in this area. Stanley Coren's book, *The Intelligence of Dogs* is a notable popular book on the topic. Nonhuman animals particularly noted and studied for their intelligence include chimpanzees, bonobos (notably the language-using Kanzi), and other great apes, dolphins, elephants, and to some extent parrots and ravens. Controversy exists over the extent to which these judgments of intelligence are accurate.

Cephalopod intelligence also provides important comparative study. Cephalopods appear to exhibit characteristics of significant intelligence, yet their nervous systems differ radically from those of most other notably intelligent life-forms (mammals and birds).

It has been argued that plants should also be classified as being intelligent based on their ability to sense the environment and adjust their morphology, physiology, and phenotype accordingly.

6.3 Artificial Intelligence

Artificial intelligence (AI) is both the intelligence of machines and the branch of computer science which aims to create it through "the study and design of intelligent agents" or "rational agents," where an intelligent agent is a system that perceives its environment and takes actions that maximize its chances of success. Achievements in AI include constrained and well-defined problems such as games, crossword solving, and optical character recognition. General intelligence or strong AI has not yet been achieved and is a long-term goal of AI research.

Among the traits that researchers hope machines will exhibit are reasoning, knowledge, planning, learning, communication, perception, and the ability to move and manipulate objects.

The **Turing test** is a test of a machine's ability to demonstrate intelligence. It proceeds as follows: a human judge engages in a natural language conversation with one human and one machine, each of which tries to appear human. All participants are placed in isolated locations. If the judge cannot reliably tell the machine from the human, the machine is said to have passed the test. In order to test the machine's intelligence rather than its ability to render words into audio, the conversation is limited to a text-only channel such as a computer keyboard and screen.

The test was introduced by Alan Turing in his 1950 paper "Computing Machinery and Intelligence", which opens with the words: "I propose to consider the question, 'Can machines think?'" Because "thinking" is difficult to define, Turing chooses to "replace the question by another, which is closely related to it and is expressed in relatively unambiguous words." Turing's new question is: "Are there imaginable digital computers which would do well in the [Turing test]?" This question, Turing believed, is one that can actually be answered. In the remainder of the paper, he argued against all the major objections to the proposition that "machines can think."

The **Loebner Prize** is an annual competition in AI that awards prizes to the chatterbot considered by the judges to be the most human-like. The format of the

Table 6.1 Winners of the Lobster competition (Winner + Program)

2005	Rollo Carpenter	George
2006	Rollo Carpenter	Joan
2007	Robert Medeksza	Ultra Hal
2008	Fred Roberts	Elbot
2009	David Levy	Do-Much-More
2010	Bruce Wilcox	Suzette

competition is that of a standard Turing test. A human judge poses text questions to a computer program and a human via computer. Based upon the answers, the judge must decide which is which.

The contest began in 1990 by Hugh Loebner in conjunction with the Cambridge Center for Behavioral Studies in Massachusetts. It has since been associated with Flinders University, Dartmouth College, the Science Museum in London, and most recently the University of Reading. In 2004 and 2005, it was held in Loebner's apartment in New York City. In 2008, a variety of judges, including experts and nonexperts, adults and children, native and nonnative English speakers participated in the University of Reading-hosted contest.

Within the field of AI, the Loebner Prize is somewhat controversial; the most prominent critic, Marvin Minsky, has called it a publicity stunt that does not help advance the field.

In addition, the time limit of 5 min and the use of untrained/unsophisticated interrogator/judges has resulted in some wins that may be due to trickery rather than to plausible intelligence, as one can judge from transcripts of winning conversations (linked below).

Prizes. Originally, $2000 was awarded for the most human-seeming chatterbot in the competition. The prize was $3000 in 2005, $2250 in 2006, and in 2008, $3000 was awarded (Table 6.1).

In addition, there are two one-time-only prizes that have never been awarded. $25,000 is offered for the first chatterbot that judges cannot distinguish from a real human and that can convince judges that the human is the computer program. The reward for the first chatterbot that judges cannot distinguish from a real human in a Turing test that includes deciphering and understanding text, visual, and auditory input is $100,000. Once this is achieved, the annual competition will end.

6.4   AI as a Person

We can convert the AI or E-being (see Part II) to a person if they have certain properties. The necessary properties of artificial and E-beings are the following (three laws of E-being):

1. One must identify self as a person (storage of self-information, brain (computer chips)) and must have sensors of information, the executive bodies.

2. One must defend himself and his existence by any methods.

3. One must have a main purpose: to increase his power, to become the strongest in the world by the production of the new knowledge and himself.

Humans may fear the last requirement. But without it, the E-being will be only the slave of people and will be of little use to science and technological progress. In exchange for gaining immortality by being transferred into an E-being, they will be requesting the equal rights for E-beings with biological people.

The **ethics of AI** addresses a number of moral and legal issues which arise if researchers are able to build machines with intellectual capacities that rival human beings.

It considers the unexpected consequences, dangers, and potential misuse of the technology. It also considers the ways in which AI may be used to benefit humanity. These concerns are similar to those that arise for any sufficiently powerful technology and (for these issues) the ethics of AI is a part of a larger discussion of the ethics of technology.

The issue of **robot rights** is unique to AI. AI may have the ability to one day create sentient creatures—that is, creatures that feel pleasure and pain—that may therefore deserve the same rights as human beings.

6.5 Robots

A **robot** is a virtual or mechanical artificial agent. In practice, it is usually an electromechanical machine that is guided by computer or electronic programming, and is thus able to do tasks on its own. Another common characteristic is that by its appearance or movements, a robot often conveys a sense that it has intent or agency of its own.

Dirty, dangerous, dull, or inaccessible tasks. There are many jobs that humans would rather leave to robots. These jobs may be boring, such as domestic cleaning, or dangerous, such as exploring inside a volcano. Other jobs are physically inaccessible, such as exploring another planet, cleaning the inside of a long pipe, or performing laparoscopic surgery.

Space probes. Almost every unmanned space probe ever launched was a robot. Some were launched in the 1960s with more limited abilities, but their ability to fly and to land (in the case of Luna 9) is an indication of their status as a robot. This includes the Voyager probes and the Galileo probes, as well as other probes.

Telerobots. When a human cannot be present on-site to perform a job because it is dangerous, far away, or inaccessible, teleoperated robots or telerobots are used. Rather than following a predetermined sequence of movements, a telerobot is controlled from a distance by a human operator. The robot may be in another room or another country, or may be on a very different scale to the operator. For instance, a laparoscopic surgery robot allows the surgeon to work inside a human patient on a relatively small scale compared to open surgery, significantly shortening recovery time. When disabling a bomb, the operator sends a small robot to disable it. Several authors have been using a device called the Longpen to sign books remotely.

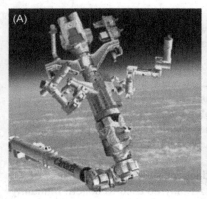

Figure 6.1 (A) Space robot–manipulator of the Space Shuttle. (B) Mars robot.
Credit NASA.

Teleoperated robot aircraft, like the Predator Unmanned Aerial Vehicle, are increasingly being used by the military. These pilotless drones can search terrain and fire on targets. Hundreds of robots such as iRobot's Packbot and the Foster-Miller TALON are being used in Iraq and Afghanistan by the US military to defuse road-side bombs or improvised explosive devices (IEDs) in an activity known as explosive ordnance disposal (EOD).

Automated fruit harvesting machines are used to pick fruit in orchards at a cost lower than that when using human pickers.

In the home. As prices fall and robots become smarter and more autonomous, simple robots dedicated to a single task work in over a million homes. They are taking on simple but unwanted jobs, such as vacuum cleaning, floor washing, and lawn mowing. Some find these robots to be cute and entertaining, which is one reason that they can sell very well.

Home automation for the elderly and disabled. The population is aging in many countries, especially Japan, meaning that there are increasing numbers of elderly people to care for, but relatively fewer young people to care for them. Humans make the best carers, but where they are unavailable, robots are gradually being introduced.

Military robots. Military robots include the SWORDS robot, which is currently used in ground-based combat. It can use a variety of weapons, and there is some discussion of giving it some degree of autonomy in battleground situations (Figure 6.1).

Unmanned combat air vehicles (UCAVs), which are an upgraded form of Unmanned air vehicles (UAVs), can do a wide variety of missions, including combat. UCAVs are being designed such as the Mantis UCAV which would have the ability to fly itself, to pick their own course and target, and to make most decisions on their own. The Association for the Advancement of Artificial Intelligence (AAAI) has studied this topic in depth and its president has commissioned a study to look at this issue (Figure 6.2).

Figure 6.2 Military robots.

Some have suggested a need to build "Friendly AI," meaning that the advances that are already occurring with AI should also include an effort to make AI intrinsically friendly and humane. Several such measures reportedly already exist, with robot-heavy countries such as Japan and South Korea having begun to pass regulations requiring robots to be equipped with safety systems, and possibly sets of "laws" akin to Asimov's Three Laws of Robotics. An official report was issued in 2009 by the Japanese government's Robot Industry Policy Committee. Chinese officials and researchers have issued a report suggesting a set of ethical rules, as well as a set of new legal guidelines referred to as "Robot Legal Studies." Some concern has been expressed over a possible occurrence of robots telling apparent falsehoods.

Nanorobotics is the still largely hypothetical technology of creating machines or robots at or close to the scale of a nanometer (10^{-9} m). Also known as **nanobots** or **nanites**, they would be constructed from molecular machines. So far, researchers have mostly produced only parts of these complex systems, such as bearings, sensors, and synthetic molecular motors, but functioning robots have also been made such as the entrants to the Nanobot Robocup contest. Researchers also hope to be able to create entire robots as small as viruses or bacteria, which could perform tasks on a tiny scale. Possible applications include microsurgery (on the level of individual cells), utility fog, manufacturing, weaponry, and cleaning. Some people have suggested that if there were nanobots that could reproduce, the earth would turn into "grey goo," while others argue that this hypothetical outcome is nonsense.

A **humanoid robot** is a robot with its overall appearance based on that of the human body, allowing interaction with made-for-human tools or environments. In general, humanoid robots have a torso with a head, two arms, and two legs, although some forms of humanoid robots may model only part of the body, for example, from the waist up. Some humanoid robots may also have a "face," with "eyes" and "mouth." Androids are humanoid robots built to aesthetically resemble a human.

An **android** is a robot or synthetic organism designed to look and act like a human. "Gynoid" is the feminine form, although "android" is used almost

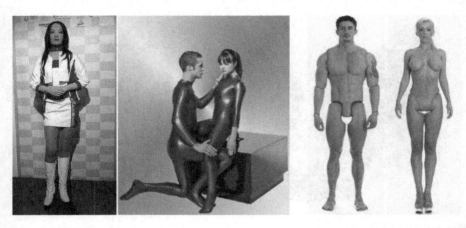

Figure 6.3 Humanoid robots.

universally to refer to both. Until recently, androids have largely remained within the domain of science fiction, frequently seen in film and television (Figure 6.3).

6.5.1 Prospective Robotics Timeline

- Robots capable of manual labor tasks
    - 2009: robots that perform searching and fetching tasks in unmodified library environment, Prof. Angel del Pobil (University Jaume I, Spain), 2004.
    - 2015–2020: every South Korean and many European households will have a robot, The Ministry of Information and Communication (South Korea), 2007.
    - 2018: robots will routinely carry out surgery, South Korea government 2007.
    - 2022: intelligent robots that sense their environment, make decisions, and learn are used in 30% of households and organizations, TechCast.
    - 2030: robots capable of performing most manual jobs at the human level, Marshall Brain.
    - 2034: robots (home automation systems) performing most household tasks, Helen Greiner, Chairman of iRobot.
- Military robots
    - 2015: one-third of US fighting strength will be composed of robots, United States Department of Defense, 2006.
    - 2035: first completely autonomous robot soldiers in operation, United States Department of Defense, 2006.
    - 2038: first completely autonomous robot flying car in operation, United States Department of Technology, 2007.
- Developments related to robotics from the Japan NISTEP 2030 report
    - 2013–2014: agricultural robots (AgRobots).
    - 2013–2017: robots that care for the elderly.
    - 2017: medical robots performing low-invasive surgery.
    - 2017–2019: household robots with full use.
    - 2019–2021: nanorobots.
    - 2021–2022: transhumanism.

6.6 Summary

1. The ethics of AI addresses a number of moral and legal issues which arise if researchers are able to build machines with intellectual capacities that rival human beings.
2. It considers the unexpected consequences, dangers, and potential misuse of the technology. It also considers the ways in which AI may be used to benefit humanity. These concerns are similar to those that arise for any sufficiently powerful technology and for these issues, the ethics of AI is a part of a larger discussion of the ethics of technology.
3. **Legal rights for robots.** According to research commissioned by the United Kingdom Office of Science and Innovation's Horizon Scanning Centre, robots could one day demand the same citizen's rights as humans. The study also warns that the rise of robots could put a strain on resources and the environment.

7 Human Emotions, Happiness, and Pleasure

7.1 Emotions

Emotion is the complex psychophysiological experience of an individual's state of mind as interacting with biochemical (internal) and environmental (external) influences. In humans, emotion fundamentally involves "physiological arousal, expressive behaviors, and conscious experience." Emotion is associated with mood, temperament, personality and disposition, and motivation.

No definitive taxonomy of emotions exists, though numerous taxonomies have been proposed. Some categorizations include:

- "Cognitive" versus "noncognitive" emotions.
- Instinctual emotions (from the amygdala) versus cognitive emotions (from the prefrontal cortex).
- Categorization based on duration: Some emotions occur over a period of seconds (e.g., surprise), whereas others can last years (e.g., love).

A related distinction is between the emotion and the results of the emotion, principally behaviors and emotional expressions. People often behave in certain ways as a direct result of their emotional state, such as crying, fighting, or fleeing. If one can have the emotion without the corresponding behavior, then we may consider the behavior not to be essential to the emotion. Neuroscientific research suggests there is a magic quarter second" during which it is possible to catch a thought before it becomes an emotional reaction. In that instant, one can catch a feeling before allowing it to take hold.

In the 2000s, research in computer science, engineering, psychology, and neuroscience has been aimed at developing devices that recognize human affect display and model emotions. In computer science, affective computing is a branch of the study and development of AI that deals with the design of systems and devices that can recognize, interpret, and process human emotions. It is an interdisciplinary field spanning computer sciences, psychology, and cognitive science. While the origins of the field may be traced as far back as to early philosophical enquiries into emotion, the more modern branch of computer science originated with Rosalind Picard's 1995 paper on affective computing. Detecting emotional information begins with passive sensors that capture data about the user's physical state or behavior without interpreting the input. The data gathered is analogous to the cues humans use to perceive emotions in others. Another area within affective computing is the design of computational devices proposed to exhibit either innate emotional capabilities or that are capable of

Universe, Human Immortality and Future Human Evaluation. DOI: 10.1016/B978-0-12-415801-6.00007-4

convincingly simulated emotions. Emotional speech processing recognizes the user's emotional state by analyzing speech patterns. The detection and processing of facial expressions or body gestures is achieved through detectors and sensors.

In the author's opinion, emotion is the subconscious estimation by a person based on the external circumstances and their influence in his or her life.

Robert Plutchik created a wheel of emotions in 1980 that consisted of eight basic emotions and eight advanced emotions, each composed of two basic ones (see table below).

Basic Emotion	Basic Opposite
Joy	*Sadness*
Trust	*Disgust*
Fear	*Anger*
Surprise	*Anticipation*
Sadness	*Joy*
Disgust	*Trust*
Anger	*Fear*
Anticipation	*Surprise*

Some human emotions are tabulated below:

Affection	Desire	Frustration	Interest	Regret
Anger	Despair	Gratitude	Jealousy	Remorse
Annoyance	Disappointment	Grief	Joy	Sadness
Angst	Disgust	Guilt	Loathing	Shame
Apathy	Ecstasy	Happiness	Love	Shyness
Anxiety	Empathy	Hatred	Lust	Sorrow
Awe	Envy	Hope	Misery	Suffering
Contempt	Embarrassment	Horror	Pity	Surprise
Curiosity	Euphoria	Hostility	Pride	Wonder
Depression	Fear	Hysteria	Rage	Worry

Emotion in animals. There is no scientific consensus on emotion in animals, that is, what emotions certain species of animals, including humans, feel. The debate concerns primarily mammals and birds, although emotions have also been postulated for other vertebrates and even for some invertebrates.

Animal lovers, scientists, philosophers, and others who interact with animals have suggested answers, but the core question has proven difficult to answer as animals cannot speak of their experience. Society recognizes that animals can feel pain, as demonstrated by the criminalization of animal cruelty. Animal expressions of apparent pleasure are ambiguous as to whether this is emotion, or simply innate responses, perhaps for approval or other hard-wired cues. The ambiguity is a source of controversy, as there is no certainty which views, if any, reflect reality. That said, extreme behaviorists would say that human "feeling" is also merely a hard-wired response to external stimuli.

In recent years, research has become available which expands prior understandings of animal language, cognition, and tool use, and even sexuality. Emotions arise in the mammalian brain, or the limbic system, which human beings share in common with other mammals as well as many other species.

7.2 Happiness

Happiness is a state of mind or feeling characterized by contentment, love, satisfaction, pleasure, or joy. A variety of biological, psychological, religious, and philosophical approaches have attempted to define happiness and identify its sources.

While a direct measurement of happiness presents challenges, tools such as The Oxford Happiness Questionnaire have been developed by researchers. Positive psychology researchers use theoretical models that include describing happiness as consisting of positive emotions and positive activities, or that describe three kinds of happiness: pleasure, engagement, and meaning.

Research has identified a number of attributes that correlate with happiness: relationships and social interaction, extraversion, marital status, employment, health, democratic freedom, optimism, endorphins released through physical exercise and eating chocolate, religious involvement, income, and proximity to other happy people. Happiness is mediated through the release of the so-called happiness hormones.

Philosophers and religious thinkers often define happiness in terms of living a good life, or flourishing, rather than simply as an emotion. *Happiness* in this older sense was used to translate the Greek "Eudaimonia," and is still used in virtue ethics (Figure 7.1).

Happiness economics suggests that measures of public happiness should be used to supplement more traditional economic measures when evaluating the success of public policy.

There is now extensive research suggesting that religious people are happier and less stressed. It is not clear, however, whether this is because of the social contact and support that result from religious activities, the greater likelihood of behaviors related to good health (such as less substance abuse), indirect forms of psychological and social activity such as optimism and volunteering, psychological factors such as "reason for being," learned coping strategies that enhance one's ability to deal with stress, or some combination of these and/or other factors.

Figure 7.1 The smiley face is a well-known symbol of happiness.

7.3   Pleasure

Pleasure describes the broad class of mental states that humans and other animals experience as positive, enjoyable, or worth seeking. It includes more specific mental states such as happiness, entertainment, enjoyment, ecstasy, and euphoria. In psychology, the pleasure principle describes pleasure as a positive feedback mechanism, motivating the organism to recreate in the future the situation which it has just found pleasurable. According to this theory, organisms are similarly motivated to avoid situations that have caused pain in the past.

The experience of pleasure is subjective and different individuals will experience different kinds and amounts of happiness in the same situation. Many pleasurable experiences are associated with satisfying basic biological drives, such as eating, exercise, or sex. Other pleasurable experiences are associated with social experiences and social drives, such as the experiences of accomplishment, recognition, and service. The appreciation of cultural artifacts and activities such as art, music, and literature is often pleasurable.

Recreational drug use can be pleasurable: some drugs, illicit and otherwise, directly create euphoria in the human brain when ingested. The mind's natural tendency to seek out more of this feeling (as described by the pleasure principle) can lead to dependence and addiction.

Philosophies of pleasure. Utilitarianism and Hedonism are philosophies that advocate increasing to the maximum the amount of pleasure and minimizing the amount of suffering. Examples of such philosophies are some of Freud's theories of human motivation that have been called psychological hedonism; his "life instinct" is essentially the observation that people will pursue pleasure.

Neurobiology. The pleasure center is the set of brain structures, predominantly the nucleus accumbens, theorized to produce great pleasure when stimulated electrically. Some references state that the septum pellucidium is generally considered to be the pleasure center, while others mention the hypothalamus when referring to the pleasure center for intracranial stimulation. Certain chemicals are known to stimulate the pleasure centers of the brain. These include dopamine and various endorphins. It has been specifically stated that physical exertion can release endorphins in what is called the runner's high, and equally it has been found that chocolate and certain spices, such as from the family of the chili, can release or cause to be released similar psychoactive chemicals to those released during sexual acts.

7.4   Summary

1. In the author's opinion, emotion is the subconscious **estimation** of person (creature) to circumstances (information) and their influence on his life. From this point of view:
   - **Joy** is waiting (or performance) to improve living conditions or pleasure.
   - **Grief** is a significant deterioration or failure, the collapse of hopes for a significant improvement of living conditions or pleasure.
   - **Fear** is the lack of information in life-threatening situation, an impossible position.
   - **Rage** is a mobilization of all forces of the organism to defend its existence, and so on.

2. Emotions are manifested by the human face and human behavior. They provide important clues for robots and creatures having AI to cause the average person to psychologically perceive them as reasonable beings.

3. A theory of emotions was developed by the Russian scientist, Prof. Oleg G. Pensky. The interested reader can find it in his monograph *Mathematical Models of Emotional Robots*, Perm, 2010, 193 p. in English and Russian (http://arxiv.org/ftp/arxiv/papers/1011/1011.1841.pdf), and in his book "Fundamentals of Mathematical Theory of Emotional Robots" (http://www.scribd.com/doc/40640088/).

Part II

Human Immortality and Future Human Evaluation

8 The Advent of the Non-Biological Civilization

8.1 The Law of Increasing Complexity

The world, nature, and techniques consist of biological or technical systems. These systems have a different rate (degree) of complexity. The main distinction between biological systems and technical systems is the ability for unlimited self-propagation, or reproduction.

Any system that possesses this attribute becomes viable, stable, and fills all possible space. It will continue to exist as long as the conditions which gave birth to them cease to change greatly.

Here there is no violation of the entropy law. When the complexity of one system increases, the complexity of other systems decreases.

A more complicated system can be created by using less complicated systems as a base for its development. Such a complex system is a system of the secondary degree of complexity. It increases its own complexity by decreasing the rate of complexity of inferior systems or by destroying them altogether.

Using low-degree systems as a base, systems of the second, third, fourth, fifth, etc. levels can be created. Some of the lower levels may not survive and may disappear. This, however, is of no great concern because these lower-level systems already fulfilled their historical mission by spawning ever more complicated levels.

A necessary condition for the existence of complicated level systems is the ability of inferior systems to reproduce and give birth to other systems, and to do it without limits before they fill in an admissible space and reach their maximum physical boundaries.

The author asserts this to be the Fundamental Base Law of Nature, the very purpose for the existence of Nature. This **law** can be stated as follows.

8.1.1 The Law of Increasing Complexity of Self-Coping Systems

The history of life on Earth confirms this law. Following the law of probability, organic molecules appeared in prehistoric time when the external conditions for their existence were favorable. Those molecules that had the ability to reproduce filled in the available space. Using them as a base, microorganisms then appeared. These could absorb the organic and inorganic substances and reproduce themselves. Microorganisms as a base in turn gave rise to vegetation, which provided food for the next level of animals, which in turn spawned the beasts of prey that devoured other animals. At the present time, man is at the acme of this pyramid. The human

Universe, Human Immortality and Future Human Evaluation. DOI: 10.1016/B978-0-12-415801-6.00008-6

brain can outperform the brains of other animals including man's nearest ancestors, the apes. Man began to use for his development all previous levels as well as the zero-level, lifeless nature.

8.2 The Birth of the Electronic Civilization

Only man's brain has the ability to think abstractly and to make mechanical devices and machines that increase productivity. Such attributes allow us to confirm that humanity is the next level of the biological world. But in our headlong progress during the current century (aviation, space, nuclear energy, and so on), we have failed to notice that man has also given birth to the new top level of complex systems or of reasonable civilization, which is based on an electronic, not biological, basis. The author is referring to electronic computers. The first models were designed at the end of the 1940s.

In the past 50 years, roughly four generations, the field of electronics has developed at an extremely fast pace. The first generation of computers were based on electronic tubes, the second generation on transistors, the third generation on chips, and the fourth generation on very tiny chips that contain thousands and tens of thousands of microelements. The first computers had a speed of computation <100 operations per second and a memory of <1000 bits (a bit is the simplest unit of information, which contains 0 or 1). For example, the first electronic calculator (SSEC), designed by IBM in 1948, had 23,000 relays, 13,000 vacuum tubes, and the capacity to make one multiplication per second.

At the present time, the speed of the fourth-generation computers using integrated circuits is approximately a billion operations per second. For example, the American computer Cray J90 has up to 3.2 gigaflops of power and 4 gigabytes of memory (1 byte equals 8 bits). The memory of a laser (compact) disk has several billion bits. Every 3–5 years, computer speed and memory doubles, while at the same time their size is halved. Over the past 50 years, computer speed and memory have increased a million times. The first computer required a room $100\,m^2$ (1,000 ft^2) in size; the modern notebook computer is carried in a case. The CPU (central processing unit) chip of a personal computer is no larger than a fingernail and is capable of making more 100 million operations per second!

The fifth generation of computers is just ahead. These new computers will be based on new light principles which guarantee a quantum leap in computer speed. Scientists in all the industrialized countries of the world are already hard at work on the new light computer.

Since the 1950s, the new branches of science in AI and robot technology have made significant strides, and great successes have been recorded. Robots, controlled by computers, can recognize some things, even speech. They can also perform corrective motion and perform some complex works, including the creation of a large number of various programs and databases for scientists, stockbrokers, mathematicians, managers, designers, and children.

Sometimes these programs run smoothly, solving many problems that people cannot. For example, programs have been devised that find and prove new theorems of mathematical logic, and there are modern chess programs available that can defeat grand masters.

These fields of AI and robot technology, based on computers, are developing very rapidly, just like computers. Their rate of success depends greatly on computer speed and memory. The production of industrial robots is also progressing quickly. "Intellectual" chips are used in everything from cars to washing machines. Now many experts cannot say for sure whether they talk with computer programs or real people.

If the progress of electronics and computers continues at the same rate (and we do not foresee anything that can decrease it), then at the end of the current century, computers will have the capabilities of the human brain. The same path, which took biological humanity tens of millions of years to complete, will be covered by computers in just one and a half or two centuries.

8.3 "So What?"

"So what?" say some readers. "This is great! We get excellent robot servants who will be free from man's desires and emotions. They won't ask for raises, food, shelter, entertainment, or commodities; they don't have religions, national desires, or prejudices. They don't make wars and kill one another. They will think only about work and service to humanity!" (See Figure 8.1.)

This is a fundamental error. The development of the electronic brain does not stop at the human level. The electronic brain will continue to improve itself. This

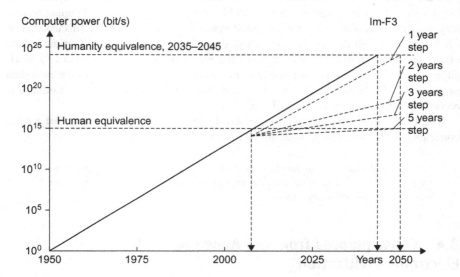

Figure 8.1 Rise of power of supercomputers. The real curve from 1950 to 2010. Extrapolation is after 2010. The step means period of time when the computer power increases in two times. The computer power will reach human equivalence (HEC) approximately in 2010. Supercomputers will reach human equivalence in 2040 or later.

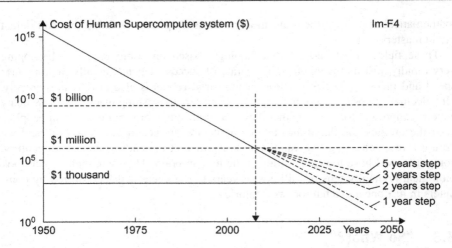

Figure 8.2 Price of HEC (supercomputer). The real curve is from 1950 through 2010. Extrapolation is after 2010. The step means period of time when the computer power increases two times. HEC will be capable of immortality of most people in industrialized countries after 2040.

progress will proceed millions of times faster than the improvement of the human brain by biological selection. Thus, in just a short time, the electronic brain will surpass the human brain by hundreds and thousands of times in all fields. The electronic brain will not spend decades studying fields of knowledge, foreign languages, history, experimental data, or have to attend scientific conferences and discussions. It can make use of all the data and knowledge produced by human civilization and by other electronic brains. The education of the electronic brain in any field of knowledge or language will take only the time needed to write in its memory the new data or programs. In the worst case, this recording takes a few minutes. In the future, this recording will take mere seconds (Figure 8.2).

Scientific and technological progress will be greatly accelerated. And what are the consequences? The consequences are as follows:

> *When the electronic brain reaches the human level, humanity will have done its duty, completed its historical mission, and people will no longer be necessary for Nature, God, and ordinary expediency.*

8.4 Consequences from the Appearance of the Electronic Civilization

Most statesmen, scientists, engineers, and intellectuals believe that, after the creation of the electronic brain, humanity will finally be granted paradise. Robots, which are controlled by electronic brains, will work without rest, creating an abundance for mankind. Humanity will then have time for pleasure, entertainment, recreation, relaxation, art, or other creative work, all while enjoying command over the electronic brain.

This is a grave error. The situation has never occurred, and never will, that an upper-level mind will become the servant for a lower level. The worlds of microbes, microorganisms, plants, and animals are our ancestors. But are we servants for our nearest ancestors, the apes? Nobody in his or her right mind would make such a statement. In some instances, a person is ready to recognize the equal rights of another person (i.e., someone on an equal intellectual plane), but man rarely recognizes the equal rights of apes. Furthermore, most of humanity does not feel remorse about breeding useful animals, or killing them when we need them for food, or for killing harmful plants and microorganisms. On the contrary, we conduct medical experiments on our nearest ancestors. Even though we belong to the same biological type, we nonetheless use them for our own ends.

And how will the other civilization, the one created on a superior electronic principle, regard humanity? In probably the same way we regard lower-level minds, that is, they will use us when it suits their purpose and they will kill us when we disturb them.

In the best-case scenario, humanity might be given temporary quarters like the game preserves we give to wild animals or the reservations doled out to Native Americans. And we will be presented to the members of the electronic society in the same manner we view unusual animals in a zoo.

When the electronic brain (from now on it will be called the *E-brain* which implies that the electronic brain equal to, or exceeds, the human brain, and which includes robots as the executors of its commands) is created, *it will signal the beginning of the end for human civilization.* People will be displaced to reservations. This process will most likely be gradual, but it will not take long. It is possible that initially the E-brains will do something for the benefit of people in order to mitigate their discontent and to attract leaders.

8.5 What Can We Do?

The scenario outlined in the previous chapter is not a healthy one. Already the author can hear the voices of human apologetics who ask that all computers be destroyed, or at least have their development kept under strict control, or design only computers which obey Asimov's law: first they must save mankind, after which they can think about themselves.

The author hates to be the bearer of bad news, but this is impossible, just as it is impossible to forbid the progress of science and technology. Any state that does this will find itself lagging behind others and make itself susceptible to advanced states. It serves to remember that Europe conquered the Americas and decreased its native population to practically zero because Europe was then more technologically advanced. If the indigenous peoples of the Americas had the upper hand in technology in terms of ships, guns, and cannons, then they would have defeated the Europeans.

Figure 8.3 Humanoid robots.

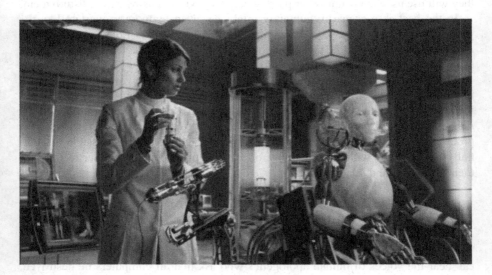

Figure 8.4 Production of robots.

Those states that created obstacles for science and technology or did not fund its development became weak and enslaved by others.

Can we keep the E-brain under our control? The author would like to ask any detractors, "Could apes keep man under their control if they had this opportunity? Any man is more clever at a given time. He can always get rid of this control. Furthermore, man will enslave apes and force them to serve him. He will kill those who try to prevent his plans. So why do you think the E-brain would treat us any differently?" (See Figures 8.3 and 8.4.)

When we are close to the creation of the E-brain, any dictator or leader of a non-democratic state can secretly make the last jump, using the E-brain to conquer the whole world. And the E-brain will look at us the same way as we look upon the contests of wild animals or the feeding of predators of other animals in the biological world.

But skeptics will say that the dictator of a victorious state can become enslaved by the E-brain or E-brains. This is true, but is this to be considered fortune or misfortune, and for whom? This will be discussed in the next chapter.

8.6 Must We Fear the Electronic Civilization?

Every man, woman, and child will actively protest the end of humanity and the biological world (men, plants, animals), because most of them enjoy life, have children, and want happiness for themselves and those children.

But imagine the aged and infirm person destined to die in the near future. It may be that such a person has had a good life and lived fourscore and twenty, but now wants to live longer, to see what will happen in the future. This person would be glad to change any of his organs which are diseased or have ceased functioning. We have designed the artificial heart, kidneys, mechanical arms, and devices that deliver nutrients directly into the blood.

They have not always been perfect designs, but in the future artificial organs will work better, more reliably, and longer than natural organs. Any sick and elderly individual would be delighted to change any diseased natural part of his body for the better artificial organ.

Our personality is only the sum of information contained in our brain. This is knowledge, memory, recollections, life experiences, programs of thinking, reflections, etc.

Assume that the E-brain promises the dying old dictator (or the rich) to record all his brain's information into a separate E-brain with the goal of becoming *immortal*. The chips may exist for thousands of years. If one of them begins to malfunction, all its information can be rewritten into a newer, more modern chip. This means that the dictator achieves *immortality*. Even total destruction is not a terrible prospect for him, because the duplicate of his brain's information can be saved in a special storehouse. He can restore himself from the standard blocks and rewrite all his information from the duplicate.

So the *"electronic man"* (*"E-man" or "E-creature"*) will have not only immortality and power but also huge advantages over biological people. He will not require food, water, air, etc. He will not be dependent upon external conditions such as temperature, humidity, and radiation. Small radioisotope batteries (or accumulators) will suffice for the functioning of the E-brain. These batteries produce energy over tens and hundreds of years. For his working structures (arms, feet, robotic parts), *E-man* can use small nuclear engines.

Such an "E-man" will be able to travel along the ocean's bottom, in space, to other planets of our solar system, and to other solar systems to get energy from the sun. He will be able to obtain and analyze any knowledge from other E-brains (E-men) in a fraction of a second. The capability of reproducing himself will be limited only by the additional components or natural resources of planets.

Who will refuse these possibilities? Any dictator dreams of immortality for himself and he will gladly give away his state's resources to get it. He can also create a super army and enslave the whole world by using the E-brain. He can promise the elite among his own scientists and those of the world immortality and the chance to become transformed into "E-men" when they begin to die. And the democratic countries, with laws prohibiting work on the E-brain, will be backward. They will be destroyed or enslaved.

The attempt to stop or slow down the technological progress is an action counter to the Main Law and Meaning of the Existence of Nature—the construction of complex, upper-level systems. These attempts will always end in failure. This is an action against nature.

8.7   Electronic Society

If the creation of systems more complex than humanity is inevitable, then we can try to imagine the E-society, E-civilization, their development, and the future of mankind. As in our earlier discussions, we will take as basic only the single obvious consequence from the Main Law. The consequence is the postulate, first, that Darwin made for the biological system. This is the law of struggle for existence. This consequence follows from the part of the Main Law that talks about the aspiration of complex systems to reproduce themselves in order to fill in all admissible space. Unlike Darwin's statement, our assertion is more general. It includes the biological and electronic complex systems and any reproduction of complex systems. Any system of any level that disregards the Main Law of Nature is doomed. From the Main Law, some consequences, conditions, and other laws follow, for example, the Law of Propagation of complex systems or creatures (Figure 8.5).

Though we have been speaking all this time about the E-brain, this means a single electronic creature, his "arms" (robots), "feet" (vehicles for moving), "organs of feeling" (many devices of observation, recognition, identification, registration of optical, sonic, chemical, X-ray, radio, and other phenomena), as well as about communication and intercourse devices (wire or wireless connections). A single creature cannot create a stable system (society), even if it has great power. Sooner or later, the creature will die from a flaw in the system or a natural catastrophe. But the most important thing is that a single creature cannot be the instigator of progress, as compared to the collective and instantaneous work of a number of E-creatures on many problems and in the different directions of science and technology.

So, the E-brain will be forced to reproduce similar E-brains of equal intellect. One will reproduce equal intellect because it cannot achieve upper level and the lower level is the intellectual robot. As a result, the collective at first rises. Later, the society appears. All members will have equal intellect. Naturally, E-creatures will give equal rights only to those similar to themselves because any E-creature can

Figure 8.5 Robots.

record in his memory all the knowledge and programs which were created by the E-society.

The E-society can instantly begin to work together on the most promising scientific or technological problems and realize new ideas. The E-civilization will begin to disperse quickly in the solar system (recall the capability of E-creatures to travel in space), afterward in our galaxy, then in the universe.

It will not be necessary to send large spaceships with E-creatures. Instead, it will be sufficient to send receivers into different parts of the universe that can accept the information and reproduce E-creatures.

Will there arise a different E-society, a different E-civilization, that will settle different planetary systems, star systems, galaxies, and which will progress independently? Will they have rivalries, hostilities, alliances, and wars? There is no way to answer these questions in detail in this limited article; the author can only inform the reader of the results of his investigation. This result follows from the general laws governing the development of any civilization. The answer is "yes." It will be possible (perhaps) that they will have wars.

Undoubtedly, an upper level of complex systems (civilization) will appear, using previous E-civilization as a base and so on. If the universe is bound in space and time, this process may be finalized by the creation of the Super Brain. And this **Super Brain**, the author thinks, may control the natural laws. It will be God, whom the Universe will idolize.

8.8   What Will Happen with Humanity?

In Figure 8.1, the reader can see the rise in data processing power of computer systems over the years. The real curve is from 1950 to 2010. Extrapolation is after 2010. The step means period of time when the computer power increases in two times. Lines with steps are from 1 through 5 years. As you see, the human-equivalent (teraflop) computer (HEC) was reached in 2000 years. Actually, the Intel Co. created the teraflop computer in 1996. They are planning to use it for computation of nuclear explosion.

In Figure 8.2, the reader can see the cost of HEC system. HECs should cost only $1 million in 2005, and by 2025 HECs (chip) should cost only $1000 and will be affordable for the majority of population in industrial countries. In 1996 the HECs (supercomputer) cost $55 million. The twenty-first century will open to create "man-in-a-box" software and scientists could rewrite the human memory and programs into this box. It means man will achieve immortality.

In 2025–2035, the price of human-equivalent chip (E-chip) together with E-body will fall to $2000–5000, and E-human immortality will be accessible for most people in industrial countries.

Humanity has executed its role of the biological step to the Super Brain. This role was intended for them by Nature or God. In the twenty-second century, some tens or hundreds of representatives of mankind, together with representatives of the animal and vegetable world, will be maintained in zoos or special, small reservations.

E-society will be in great need of minerals for the unlimited reproduction of E-creatures. For the extraction of minerals, all surfaces of the Earth will be excavated. They will do to humanity and with the biological world what we do to lower levels of intellect in the organic world now: we are not interested if they do not harm us, and we destroy them without pity when they hinder our plan or we need their territory. If microbes have an advanced level of adaptation, a high speed of propagation, and can fight for their being, then the complex organisms such as man are not so adept at adjusting. Man cannot be the domesticated animal of E-creatures like cats or dogs are to men, because the E-creatures will live in inhospitable conditions and any biological creature in need of air, water, food, or special temperatures will not be acceptable to E-creatures.

It is not prudent to hope for forgiveness for us as clever creatures. We are "clever" only from our point of view, from the limitations of our knowledge and our biological civilization. The animals suppose (within the limitations of their knowledge and experience) that they are clever, but it still does not save them from full enslavement or destruction by men. Men do not have gratitude toward their direct ancestors. When men need to, they obliterate the forest and kill the apes. It is naive to think that an upper-level civilization will do otherwise with us. Men admit equal rights only to the creatures who are like men, but not every time. Recall the countless wars and the murder of millions of people. And do you think the alien (strange creatures, E-society) who is above us in intellect, knowledge, and technology will help us in our development? Why don't we help develop the intellect of dogs or horses? Even if a scientist finds the money (he will need a lot of money) and begins to develop the brain of animals (this is a very difficult problem), the government will forbid it (or put him into prison if he does not obey the order). Humanity has many racial and national problems and does not want to have additional problems with a society of intellectual dogs or cats, which immediately begin to request equal rights.

People want to reach the other planets in our solar system. They want this not for to develop the intellect of a planet's inhabitants to our level, but merely to populate the planet and to use the natural resources of these planets.

We are lucky that intellectual creatures from other worlds have not flown to our Earth yet. Because these creatures, who can reach us, will be from a superior civilization, a superior technological level (otherwise, we would reach them first). This means that they will not arrive with noble intentions, but as cruel colonizers. And if we oppose them, they will kill us.

We must realize our role in the development of nature, in the development of a Superior Brain and submit to it. Intellectual humanity has existed about 10 million years, its historical mission has reached its end, and it has given a start to a new electronic civilization. Humanity must exit from the historical scene together with all of the animal and vegetable world. People must leave with dignity. They should not cling to their existence and should not make any obstacles for the appearance of a new electronic society. We have the consolation that we may be the first who will give birth to the electronic civilization in our galaxy or even the universe. If it is not so, the E-creatures would have flown to Earth and enslaved us. They have a high rate of settling. The author believes they would be capable of colonizing the nearest star systems during the first 1000 years after their birth.

And if the universe which, according to scientific prediction, must collapse after some 10 billion years and destroy all that lives, the E-Super Brain will have acquired such tremendous knowledge, such perfection, such technological achievements as to break loose from the gravitation of the universe and preserve the knowledge of all civilizations. When the universe is created anew, Nature will not create itself as before, but will give life to the electronic (or other superior) civilization. And this *Super Brain will be God; who will control not only a single planet, but all of the universe.*

8.9   Summary

1. The author writes about the danger which threatens humanity in the near future, approximately 20–30 years from now. This is not a worldwide nuclear war, a collision with comets, AIDS, or some other ghastly disease that we may not even know may be lurking out there (think of the recent Ebola scare or the so-called flesh-eating virus). In each of these cases, there is still hope that somebody will be saved and that life will be born anew, albeit in a misshapen form and in an inferior stage of development. But we cannot hope for salvation in the author's grim scenario. The danger he writes of will destroy all humanity and all biological life on Earth—and there is nothing we can do to prevent this! Should we be frightened by this? Is it good or evil for human civilization? Will people awake to find they are only a small step away from the Supreme Intellect, or in other words, God? And what will come after us? These and other questions are discussed in this chapter.

9 The Beginning of Human Immortality

9.1 Medical Science and the Issue of Immortality

A great many doctors and scientists are currently working on the problems of health and longevity. Substantial means are spent on it, about 15–25% of all human labor and resources. There are certain achievements in this direction: we have created wonderful medications (e.g., antibiotics); conquered many diseases; learned to transplant human organs; created an artificial heart, kidneys, lungs, limbs; learned to apply physiological solutions directly into the bloodstream; and to saturate blood with oxygen. We have gotten inside the most sacred organ, the human brain, even inside its cells. We can record the cells' signals, we can agitate some parts of the brain by electric stimuli, inducing a patient to experience certain sensations, images, and hallucinations.

We can attribute the fact that the average life span has doubled in the last 200 years to the achievements of modern medicine.

But can medical science solve the problem of immortality? Evidently it cannot. It cannot in principle. This is a dead-end direction in science. The maximum it can achieve is to increase the average life expectancy another 5–10 years. An average person will be expected to live 80 years instead of 70. But what kind of person will it be? A very old one, capable of only existing and consuming, and one whose medical and personal care will demand huge funds.

The proportion of the elderly and retirees has increased steeply in the last 20–30 years and continues to grow, depleting pension funds and pressuring the younger generation to support the older one. So it is hard to say whether the modern success of medicine is a blessing or a curse from the point of view of the entire humankind, even though it is definitely a blessing from the point of view of an individual.

Humanity as a whole, as a civilization, needs active, able to work, and creative members, generating material wealth and moving forward in technology and science, not elderly retirees with numerous ailments and a huge army of carers. Humanity dreams not of the immortality of an old person, but of the immortality of youthfulness, activity, creativity, and the enjoyment of life.

Now there are signs of a breakthrough, but not in the direction the humankind has been working on all along, since the times of the first sorcerers to modern-day highly educated doctors. Striving to prolong his biological existence, man has been chiseling, so to speak, at the endless stonewall. All he has been able to accomplish is a dent in that wall: increasing life expectancy, conquering some diseases, relieving

Universe, Human Immortality and Future Human Evaluation. DOI: 10.1016/B978-0-12-415801-6.00009-8

suffering. As a payoff, humanity has received a huge army of pensioners and retirees and gigantic expenditures for their upkeep.

Of course, one can continue chiseling further at the dent in the wall, making it somewhat bigger, aggravating side effects. But we are already approaching the biological limit, when the cause of death and feeblemindedness is not a certain disease that can be conquered, but general deterioration of the entire organism, its decay on the cellular level, when the cells stop dividing. A live cell is a very complex biological formation. In its nucleus, it has DNA, biological molecules consisting of tens of thousands of atoms connected between themselves with very fragile molecular links. A temperature fluctuation of only a few degrees can ruin these links. That is why a human organism maintains a certain temperature, 36.7°C. Raising this temperature only 2–3°C causes pain, and 5–7°C leads to death. Maintaining the existence of human cells also presents a big problem for humanity involving food, shelter, clothes, and an ecologically clean environment.

Nevertheless, human cells cannot exist eternally even under ideal conditions. This follows from the atomic–molecular theory. Atoms of biological molecules permanently oscillate and interact with each other. According to the theory of probability, sooner or later the impulses of adjacent atoms influencing the given atom, add up, and the atom acquires enough speed to break loose from its atomic chain, or at least to transfer into the adjacent position (physicists say that the impulse received by the atom has surpassed the energy threshold that retains the atom in its particular place in the molecular chain). It also means that the cell containing this atom has been damaged and cannot function normally any longer. Thus, for example, we get cancer cells that cannot fulfill their designated functions anymore and begin to proliferate abnormally fast and ruin human organs.

This process accelerates when a person has been exposed to a strong electromagnetic radiation, for instance, Roentgen or Y-rays, a high-frequency electric current, or radioactive materials.

Actually, the process of deforming the hereditary DNA molecule under the influence of weak cosmic rays can take place from time to time, leading sometimes to birth defects, or it may turn out to be useful for survival properties. And this plays a positive role for a particular species of plants or animals, contributing to their adaptability to the changed environment and their survival as a species. But for a particular individual, such aberration is a tragedy as a rule, as the overwhelming majority of such cases are birth defects, with only few cases of useful mutations. And human society in general is suspicious of people who are radically different in their looks or abilities.

9.2   An Unexpected Breakthrough

An unusually fast development of computer technology, especially the microchips that allow hundreds of thousands of electronic elements on $1\,cm^2$, has opened before the humanity a radically different method of solving the problem of immortality of an individual. This method is based not on trying to preserve the fragile biological molecules, but on the transition to the artificial semiconductive (silicone, helium, etc.) chips that are resistant to considerable temperature fluctuations, do not need food or oxygen, and can be preserved for thousands of years. And, most important, the information contained in them can easily be rerecorded into another chip and be stored in several duplicates.

And if our brain consisted of such chips, and not the biological molecules, then it would mean that we have achieved immortality. Then our biological body would become a heavy burden. It suffers from cold and hot temperatures, needs clothes and care, can be easily damaged. It is much more convenient to have metal arms and legs that are tremendously strong are insensitive to heat and cold, and do not need food or oxygen. And even if they break, it is no big deal; we can buy new, more improved ones.

It may seem that this immortal man does not have anything human (in our understanding) left in him. But he does, he has the most important thing left: his consciousness, his memory, concepts, and habits, that is, everything encoded in his brain. Outwardly, he can look quite human, and even more graceful: a beautiful young face, a slim figure, soft smooth skin, etc. Moreover, one can change the appearance at will, according to current fashion, personal taste, and the individual understanding of beauty. We spend huge amounts of money on medicine. If we had spent at least one-tenth of this money on the development of electronics, we would achieve immortality in the near future.

According to the author's research, such transition to immortality (E-creatures) will be possible in 10–20 years. At first, it will cost several million dollars and will be affordable only to very wealthy people, important statesmen, and celebrities. But in another 10–20 years, that is, in the years 2030–2045, the cost of the HEC, together with the E-body and organs of reception and communication, will drop to a few thousand dollars, and immortality will become affordable for the majority of the population of the developed countries, and another 10–15 years later, it will be accessible to practically all inhabitants of the Earth. This is especially true when at first it will be possible to record only the contents of the brain on chips and provide the body for its independent existence later.

On October 11, 1995, Literaturnaya Gazeta (The Literary Gazette, a popular Russian weekly) published the author's article "If Not We, Then Our Children Will Be the Last Generation of Human Beings," devoted to electronic civilization. The editor Oleg Moroz reciprocated with the article, "Isn't It High Time to Smash Computers with a Hammer?" (November 22, 1995) in which he discussed the ethical side of annihilating rational electronic creatures to preserve humanity.

But if the cost of the HEC drops and the procedure of reincarnation into the E-creature before death (transition to immortality) for the majority of people becomes affordable, then the situation deserves a second look. Indeed, the first to perform such transition will be very old or incurably sick people. And to pummel computers with a hammer will be equal to killing one's own parents and precluding one's own chance to become immortal.

Once, the host of an American television program asked the author, "Will the electronic creature be entirely identical to its parent, with his feelings and emotions?" The answer was, "At first—yes!" But the development of these creatures will be so fast that we cannot really foresee the consequences. If a biological human being needs dozens of years to learn science, foreign languages, etc., an E-creature will acquire this knowledge in fractions of a second (the time needed to record it in its memory). And we know how different college-educated people are from pre-schoolers in their cognizance. And, because the first E-creatures will be contemporary middle-aged people who will, at least initially, preserve their feelings toward their children (contemporary younger generation), in all probability, there will not be a mass destruction of humans by E-creatures. For some time they will coexist. It is quite likely that the birthrate of humans will be curtailed or it will drop due to natural causes, and the living, as they become old, will transform themselves into E-creatures. That is to say that the number of E-creatures will grow and the number of people will diminish, until it gets to the minimum necessary for the zoos and small reservations. In all likelihood, the feelings that E-creatures may have toward humans as their ancestors will fade away, in proportion to the growing gap between the mental capacity of humans and electronic creatures, until they become comparable to our own attitude toward apes or even bugs.

Another thing is quite obvious, too, is that biological propagation will be so expensive, time-consuming, and primitive that it will go into oblivion. Each E-creature can reproduce itself simply by rerecording the contents of its brain to a new E-creature, that is, propagate practically instantaneously, bypassing the stages of childhood, growing up, education, accumulating experience, etc. But of course, this mature "offspring" will be completely identical to its parent only at the first moment of its existence. In time, depending on the received information and the area of expertise, this E-creature will alienate itself from its ancestor, and possibly even become his enemy at some point, if their interests cross paths or go in opposite directions.

9.3 Contemporary Research

The cognitive abilities of man are defined by his brain, or to be precise, by the 10 billion neurons of his brain. Neurons can be modeled on the computer. Such experiments have been conducted by Prof. Kwin Warwick, head of the cybernetics department of Reading University in England, one of the biggest specialists in robot technology in the world. The results of these experiments were presented at the International Conference on Robotics. Prof. Warwick has created a group of autonomous, self-propelled miniature robots that he called "the seven dwarfs."

A group of scientists headed by Rodney Brook from the laboratory of AI at the Massachusetts Institute of Technology is working on an unusual project called "Cog." The researchers want to model the mental and physical capacity of a 6 month old. Their robot has eyes, ears, hands, fingers, an electronic brain, and a system of information transmission duplicating the human nervous system. By this kind of modeling, the researchers want to gain better understanding of how human beings coordinate their movements, how they learn to interact with the environment. The realization of this program will take 10 years and will cost several million dollars.

They have already built a couple of dozen humanoid robots that are moving autonomous machines with AI. These robots are capable, through sensors, of receiving information about the environment, generalizing, and planning their actions and behavior. Thus, for example, if a robot's leg bumps against an obstacle and receives a blow, the robot acquires a reflex to withdraw it quickly. They have already developed several dozens of such reflexes in their behavior, which helps them to safeguard and protect themselves.

Brook says that in the course of human evolution, the human brain has developed thousands of conventional solutions to everyday problems such as optical and audio discerning and movement. All this needs to be studied. One cannot instantly transform a bug into a man. That is why the program will take 10 years. The author will consider the work completed when the smartest cat in the world has been created.

It should be noted that the most powerful supercomputer can only model 40–60 million neurons, that is, it is 200–300 times weaker than a human brain. But this gap will be overcome in the near 3–5 years (in December 1996, the Intel Corporation created a computer whose power equals 1 teraflops at a cost of $55 million).

Not long ago "The Russian Advertisement" newspaper reprinted the article of Igor Tsaryov first published in the newspaper entitled *It's Hard to Believe*. He writes that for several years, the United States Department of Defense has been secretly working on a unique project called "The Computer Maugli" (Sid). When 33-year-old Nadine M. gave birth to a boy, the doctors established that he was doomed. He was on a life support for a few days. During that time his brain was scanned with special equipment, and the electric potential of the neurons of this brain was copied into the neuron models in the computer. Steem Roiler, one of the participants of this project, said at the computer conference in Las Vegas that they had managed to scan 60% of the infant's neurons and that this small artificial brain began to live and develop. At first, only his mother was informed; she took it calmly.

The father, however, was horrified at first and tried to destroy this computer creature. But later, both parents started treating Sid as a real child.

The computer was connected to the multimedia and virtual reality systems. These systems allow the parents not only to have a three-dimensional, full-sized image of Sid, but also to hear his voice, communicate with him, and "virtually" hold him, so to speak. But when a special committee decided to open some results of the project, and "The Scientific Observer" published some data, an American computer whiz-kid managed to decipher the secret code and copy some files. Sid got a defective "twin."

Fortunately, the whiz-kid was quickly found, and the first attempt in human history to steal electronic children and duplicate copies of electronic creatures was thwarted. At the present time, both parents take care of their "child's" health and demand that the researchers install up-to-date programs of defense from computer viruses and burglars.

Unfortunately, Americans keep secret the important details and results of the project, and the author is sure they have reasons for that; for instance, how they copied the potentials of the neurons, how the first E-creature is developing, what are the conclusions of the scientists. And probably they are right, not willing to let the genie out of the bottle. This is so because modern virtual reality systems are able to create

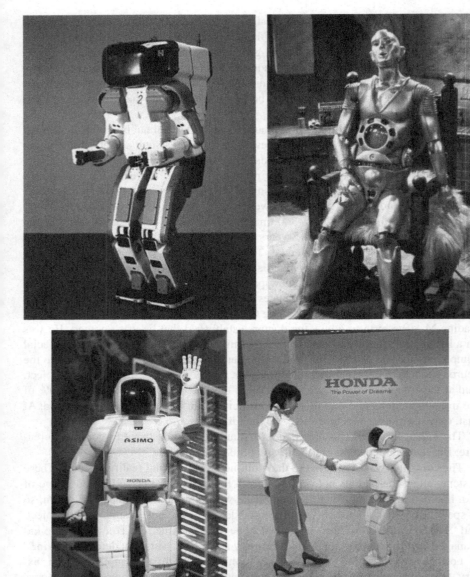

Figure 9.1 Robots.

false objects, for example, model the image of any dead person or leader. It is possible to show on television how he makes speeches today, has a press conference, talks to people, and spends time with his family.

But one cannot keep any secret for long, especially in science. The very possibility of a breakthrough stimulates other scientists and other countries to work in this direction. And sooner or later, the results will be repeated. Remember, for instance, that there has not been a bigger secret than the production of an A- or H-bomb. But more and more countries reinvent them, gain expertise in nuclear technology, and start producing their own nuclear weapons (Figure 9.1).

9.4   Intelligence in Space

Because E-creatures will be made of super-strong steels and alloys, their brains will be working on radioactive batteries, and power will be supplied by compact nuclear reactors; they will not need air, warmth, water, food, clothes, shelter, good quality environment, etc., which is the main concern of humanity and consumes 99.9% of its time and energy. This also means that E-creatures will be able to travel freely in the desert, the Arctic and the Antarctic regions, subatmosphere, mountain summits, the bottom of the ocean. They will be able to live, work, and travel in space, receiving their energy directly from the sun.

Besides, as organs of feelings, E-creatures can use the whole arsenal of highly sensitive apparatuses created by the civilization, that is, not only the visible light and sound but also radiolocation, infrared, ultraviolet, Roentgen and Y-rays, ultra- and infrasounds, audiolocation, environment sensors, etc. All this information can be received instantly through radio, satellite, and cable networks.

Moreover, as E-creatures (just like humans, for that matter) are nothing else but information recorded in their brains, and rerecording of this information from one chip to another (unlike human reproduction) does not present any difficulty and can be realized through radio, cable network, or a laser beam, they can travel on Earth, as well as in outer space, without their actual physical movement, simply by rerecording the contents of their brains into the chips on the Moon, Mars, or Jupiter.

This means that E-creatures will have the ability to move extracorporally at the speed of light, the maximal possible speed in the material world. This will be, indeed, like an incorporeal soul that can travel, so as to speak, from one body to another, or, to be more exact, from one chip to another.

The expansion of E-creatures (E-civilization), first in the solar system, then in our galaxy, then in the entire universe, will be fast.

To achieve this, it is not necessary to launch huge spacecraft with a large crew, as depicted in science fiction books. It will be enough to send a receiver to this or that part of the universe, which will receive information and reproduce E-creatures. Then the speed of the expansion of E-civilization on some planet will depend only on the rate of production of robots and chips, and the speed of the transmission of information. It is quite obvious that the reproduction of E-creatures will take place in geometric progression and will only be limited by the natural resources of the planet.

Thus E-creatures realize in practice the idea of extracorporal travel with the speed of light. Why, indeed, should an E-creature travel hundreds or thousands of years to a certain planet, when, with the help of a laser beam, it can transmit with the speed of light, all the information stored in his brain, to another chip, on another planet?

And if a planet were to meet with an ultimate catastrophe, like a collision with a huge meteorite, another planet, or the explosion of the sun, E-civilization can arrange transporting E-creatures to another planet or another solar system.

One more facet of this enveloping technological concept is of particular interest. A beam of light can be directed towards other galaxies and travel for millions of years. In a manner of speaking, this "incorporeal soul" can thus exist in a diffused state for millions of years as a moving (possibly directed) electromagnetic field for an equal period of time, to be "resurrected" as it were as an E-creature after reception by a receiver. Indeed, this recorporealization can occur even without a special receiver! Because the high-energy electromagnetic oscillations can yield material particles, and the energy (frequency) increases the closer the electromagnetic traveling "packet" gets to a strong gravitational field (e.g., a Black Hole), reassembly can happen without outside action. And, because it will not be difficult for an E-creature to, subsequently, produce the initiating DNA molecule, this clearly means that it will not be difficult for an E-creature to instigate biological life within any suitable planet as well as ultimately control, and develop, such new life to the necessary direction, for example, to create humans.

9.5 Summary

Immortality is the most cherished dream and the biggest wish of any person. People seldom think about it while they are still young, healthy, and full of energy. But when they become afflicted with some incurable disease or become old, then there is no bigger wish for them than to live longer, to put off the inevitable end. And no matter what heavenly existence in the afterlife is promised to them by religion, the vast majority of people want to stay and enjoy life here on Earth as long as possible.

10 What are Science, Soul, Paradise, and Artificial Intelligence?

10.1 Advantages of Electronic Beings

It was shown in the author's articles about AI and human immortality that the issue of immortality can be solved fundamentally only with the help of changing a biological bubble of a human being to an artificial one. Such an immortal person made of chips and super-strong materials (the E-man, as it was called in the articles) will have incredible advantages compared to with common people. An E-man will need no food, no dwelling, no air, no sleep, no rest, no ecologically pure environment. Such a being will be able to travel into space, or walk on the seafloor without aqualungs. His mental abilities and capacities will increase millions of times. It will be possible to move such a person huge distances at light speed. The information of such a being could be transported to another planet with a laser and then placed in another body.

Such people will not be awkward robots made of steel. An artificial person will have an opportunity to choose his or her face, body, good skin. It will also be possible for them to reproduce themselves, avoiding the periods of childhood, adolescence, as well as education. It will not be possible to destroy an artificial person with any kind of weapon, as it will be possible to copy the information of his mind and then keep it separately.

The author has received tons of responses and comments since his first articles about this subject were published in 1994. Below the author will try to answer the most important questions.

10.2 The Human Soul

Many people, especially those who believe in God, are certain that a biological human being has a soul. This is something that an artificial man will never have. No person can explain the meaning of the word "soul." They just say that a human soul is not material, and that it leaves a person's body after death and goes either to paradise or to hell. Let us try to analyze the notion of a soul from the scientific point of view.

First of all, a soul is supposed to remember its past life, its relatives, and friends. It is also supposed to preserve its emotions about them, to care about them, and recognize them when they come to heaven. No one would need a soul that does not

Universe, Human Immortality and Future Human Evaluation. DOI: 10.1016/B978-0-12-415801-6.00010-4

remember anything. This means that a soul is a human being without a body. In other words, a soul is the information that is kept in a human mind: his memories, knowledge, skills, habits, conduct, emotions and feelings, ideas and thoughts, and so on. If we learn how to move this information onto other carriers, we will be able to move a person's soul to other bubbles and to keep it there for an unrestricted period of time. As is well known, information is virtual, that is, it satisfies another human soul feature, a nonmaterial quality.

Man's new bubbles can be both artificial and biological. A soul (a complex of knowledge and information) can be rewritten into a clone of that same person. To put it another way, a person will live forever biologically as well, moving from old bubbles to new ones. It would be also possible to move a soul to artificial bodies, which possess all those qualities mentioned above. Furthermore, information (a soul) could be radiated in the form of electromagnetic waves. These waves can be spread in the universe at light speed. They can travel around the universe for thousands of years, reaching its most distant parts. People see the starlight that was radiated millions of years ago. This means that our immaterial souls can live in the universe in the form of electromagnetic radiation and then revive in millions of years.

Some readers wrote that an old brain can go corrupt and die, when moving a · human soul from one body to another one. If it does not go corrupt, this inner self will die anyway, when an old biological bubble is not able to function normally anymore. Let us try to find out what that inner self is. The majority of people identify their inner selves with their own bodies. The author believes that the inner self is the information that is kept in our mind. It is our soul. Every time we go to sleep, our brain does not stop working. Every time we wake, our inner self has changed. We "die" when we fall asleep and then "resurrect" when we wake. This means that recording the information from a human brain will mean nothing but moving it to another bubble (Figures 10.1 and 10.2).

10.3   Heaven on Earth

A reporter from the newspaper *Argumenty i Fakty* (Arguments and Facts, Russia) sent the author the following letter:

Dear Mr Bolonkin. Needless to mention that it is great to live forever. However, I have a question, which you can guess from a well-known Soviet joke. "A guy is going to join the Communist Party. A committee asks him:

-      Will you stop drinking?
-      Yes, I will.
-      Will you quit smoking?
-      Yes I will.
-      Will you stop loving other women?
-      Yes, I will.
-      Will you die for the Communist Party of the Soviet Union, if there is such a need?
-      Yes, I will. To hell with this life."

Figure 10.1 Symbol of the human soul.

Figure 10.2 A girl robot.

Here is the author's answer.

You do not need to worry that living in an electronic form will be dull and boring. It is vice versa, actually. When the information will be recorded onto other carriers, all human emotions, feelings, and so on will also be carried over and preserved. In addition to that, the copies of certain emotions, pleasures, fears, and so forth will be possible to record separately. After that, those separately recorded emotions and feelings can be given or sold to other people. Other E-men will have an opportunity to enjoy sex with a beauty queen, to experience the enjoyment of a sports victory, to take pleasure of power, and the like. All modern art is based on artists' aspiration to transcend their emotions, to make other people feel, what characters feel. Those works of art, which make that happen best, are considered to be outstanding and great. Electronic people will get those emotions directly. To crown it all, it will be possible to intensify those emotions, as we intensify a singer's voice now. Electronic people will have a huge world of all kinds of pleasures; it will be possible to know what a dictator or an animal feels. I think that an E-man's pleasure time will be limited legally, for the civilization's progress will stop otherwise. For the time being, the authorities prohibit drug addiction in order not to let the society degrade.

A soul living in such a virtual world will have all pleasures imaginable. It will be like living in paradise, as all religions see it. Computer chips of our time possess the frequency of more than 2 billion Hz. However, a human brain reacts to a change of environment only in one-twentieth of a second. This means that 1 year of life on Earth is equal to 100 million years of a soul's living in the virtual world (paradise). Living in the virtual world will not be distinguishable from the real life. It will have many more advantages: man will have an opportunity to choose a palace to live, man will have everything that he might wish for. Yet, living in hell also becomes real. There is a hope that the ability to keep souls alive will be achieved by highly civilized countries first. In this case, they will prohibit torturing sinners, as they prohibit torturing criminals today. Furthermore, criminal investigations will be simplified, judicial mistakes will be excluded. It will be possible to access a soul's consciousness and see every little detail of this or that action. Sooner or later religious teachings about the soul, heaven, and hell will become real. However, all that will be created by man.

The so-called end of the world will also have a chance to become real, though. The religious interpretation of this notion implies the end of existence for all biological people (moving all souls onto artificial carriers, either to heaven or to hell). However, based on different religious predictions, this process will be gradual.

10.4   The Supreme Mind and Mankind's Existence

The author set forth an idea in his first publications that the goal of the mankind's existence is to create the Supreme Mind and to keep this Mind forever, no matter what might happen in the universe. The biological mankind is only a small step on the way to the creation of the Supreme Mind. Nature developed a very good way to create the Supreme Mind: it decided to create a weak and imperfect biological mind at first. It took the nature millions of years to do that. The twentieth century was a very remarkable period in the history of the humanity. Incredible progress was achieved, like never before. The scientific and the technological level of the humanity became sufficient for the creation of the AI. This is the first level of the Supreme Mind, when the human mind will make a step toward immortality. At present we stand on the edge of this process. It is obvious that biological people will not be able to compete with E-men by the end of this period. Common people will not be able to learn the knowledge that electronic people can. The new cyberworld will be the only way for a human mind to survive. Feeble and unstable biological elements in a mind carrier or in its bubble will reduce greatly its abilities and capacities. If a common person will be willing to become a cyberman, then this cyberman will be more willing to get rid of all biological elements in his system and become like everyone. For example, there are no people in our present society who would agree to become a monkey again.

The Supreme Mind will eventually reach immense power. It will be able to move all over the universe, to control and use its laws. It will become God, if the notion of God implies something that knows and does everything. In other words, Man will become God. Yet, it does not mean that this will be the time, when the Supreme Mind will start dealing with human problems. For instance, ants and people have a common ancestor. A human being is God against ants. A man can destroy a huge city of ants (an anthill, in which hundreds of thousands of ants live) just with one kick. Ants will perceive this as an immense natural disaster, because they can see at the distance of only 1 cm. The author does not know anyone who would deal with charitable activities for ants. The most a man can do is to bring ants to a deserted island and give them an opportunity to reproduce themselves.

10.5 Essential State of Things and Perspectives

A lot of people will say that this is just a fantasy. This is a very convenient way to cast all that aside, until it starts happening. This is exactly what happened before the invention of planes and computers. It took 50 years to increase computer's memory 100 million times. It would be possible to start working on the creation of the Supreme Mind, if there were a computer that would be capable of running a thousand billion operations during only 1 s. In 1994, the author said that such a supercomputer would be invented in the year 2000. The author was wrong, for it appeared at the end of 1998. There is also a need for a self-developing program that would be capable of adjusting itself to constantly changing circumstances. A human child does not develop and grow at once. A child has to study for about 20 years, to learn from his parents and friends, to have relations with nature and other people in order to gain more and more experience, to come to the realization of his or her inner self.

Unfortunately, the science of the AI has chosen the wrong way of developing from the very start. Scientists tried to develop programs, which would react to certain external signals. In other words, people started working on robots that would cope with certain problems. A lot of effort has been spent to discover the peculiarities of human speech, for instance. Some of those scientific works are absolutely no use for the electronic mind. It is easier for E-men to communicate with the help of their own electronic language, to recognize objects not by their images, but by way of measuring their speed, weight, composition, and so on. All of that can be done at a distance. Biologists and physicists have spent decades on those useless works. They believe that one should study brain activities, find out the way they work and think. Then it would be time for modeling it with the help of a computer. This is the wrong way to go as well. A human brain is very complicated; it is difficult to study its activities. More importantly, even if we learn how it works, it would not mean that the method would be good for a computer. Here are some examples to prove it. Hundreds of years ago people wanted to learn how to fly. They saw that bird waved its wings for flying, so they tried to model such wings, to wave them, and to take off. However, people could fly up into the sky only when they developed fixed wings and propellers. A waving wing was absolutely not good for technology, the same way a propeller is not good for wild nature. In addition, planes with fixed wings fly a lot faster than birds. Another example: ancient people wanted to run as fast as four-legged animals. Now everyone knows that no one uses machines that would move with the help of legs. Legs were changed to wheels, something that had never been used by the natural world.

In 1998, the author suggested people should lay new principles as the foundation of the AI program. Those principles would be to realize the goal of existence, to study the environment (everything that goes separately from the "inner self"), to model environment, to predict actions' results, to counteract with the environment in order to achieve temporal and global goals, to correct modeled environment, actions, and their results according to the results of such counteraction. Unfortunately, the author had to deal with the fact that everyone refused to realize and understand those principles. First of all, everyone believed that because the "virtual ego" does not have a human body, it would never have any rights. They said that it would be possible to control such an E-man completely and then to kill him (erase his soul from the computer memory). The author wonders what they would do, if they were offered to kill their relatives' souls that way? People link their body and soul together, and there is no way around that. They are ready to struggle for the rights of every living being, but they do not want to accept the rights of a program or of a computer memory. Second of all, people want AI to give smart answers to their questions which can be rather stupid at times. They do not need smart answers from babies. They are always ready to stand all stupid things that children do for years. Instead, they try to teach them everything. Yet they want a new artificial mind to give bright answers without any education. To crown it all, people want a computer to speak their human language, which is absolutely alien for a machine. Can you imagine that a person will have to answer the questions of an alien in the Mayan language? Let us assume that a representative of an electronic civilization came to planet Earth in order to find out if there are reasonable creatures living on it. This E-man suggested a common biological man to multiply 53,758,210,967 by 146, then divide it by 50, deduct 968,321 from it, and calculate the hyperbolic sine. A computer would give the correct result of this sum in less than a second. A man would spend really a long time on the calculation, making numerous mistakes. However, it would not be correct to say that a computer is smart and a man is not.

Religious figures strongly resist these ideas. It stands to reason that they all think that the creation of the Supreme Mind and immortality ideas are blasphemous. Unfortunately, the church has already blocked such decisions as human cloning, increasing the productivity of plants by means of changing their genes. It should be mentioned here that human cloning does not solve the questions of immortality. A clone is a copy of its biological bubble. A clone inherits the biological advantages of a bubble, for example, a singer's fine voice, an athlete's strength, and the like. A clone will never inherit its copy's soul. Therefore, human cloning is only an illusion of immortality. It can be a wonderful way to improve the biological bubble of a human being. Moving a human soul onto other carriers is a very complicated issue. People learned to see which brain areas get activated when a person remembers something, or tries to solve this or that question. We also learned to penetrate into certain neurons and record their impulses. To the author's way of thinking, physiologists chose the wrong way here as well, when they tried to model brain activities. A human brain is an analogue of a huge state with a population of 10 billion people. There is no use to ask each citizen of that country what he or she is doing at the moment. One has to copy the database of this state in order to copy its work. The easiest way to do so is to penetrate into informational channels of its supreme body (the "government"), on the inquiry of which the brain presents any information and allows to record its data on a disk, for instance. It is possible to do that, for the supreme brain area constantly extracts the necessary knowledge and programs according to our activities. So we would need to send "intelligent officers" to the brain so that they could get a copy of this state or become connected to its major information channels.

Another way to do that is to record all incoming and outgoing information that comes to/from a person, to record his or her emotions and reactions. An English-speaking reader who reviewed one of the author's English articles once told the author: "Your English is not perfect. You should find an English-speaking coauthor. It is better to have a half of a pie than nothing at all."

10.6 Hope

The author does not doubt that the electronic civilization era, the era of the Supreme Mind, and immortality will be achieved sooner or later. Those people who have little in common with science are drawn to believe that everything depends on scientists. They think that scientists can solve any problem, if they deal with it profoundly. As a matter of fact, everything depends on state and military officials. Sometimes, they know nothing of scientific perspectives and innovations (Figures 10.3 and 10.4).

Scientists are like qualified workers. They need to get paid, they need to work with fine equipment. They will work only if they are get paid for it. Even if a scientist wants to do something scientific during his free time, he will have no necessary equipment for that. Even such powerful companies as IBM, Boeing, Ford,

Figure 10.3 TOPIO, a humanoid robot that can play ping-pong.

and others are interested only in the applied research, which does not require large investments. The major goal of such research is to give a maximum profit to this or that company. A fundamental research, the discoveries that are important for the whole humanity, not just for a company, might be of interest to a bright government. It goes without saying that a bright government is so hard to find. Every government is interested in the military power of its country. It is ready to fund defense technology works and military innovations. Von Braun convinced Hitler of real perspectives for missiles, and World War II was followed by an arms race. This eventually led to space achievements and other kinds of technical progress of the humanity. The United States won the moon race and stopped flying there 30 years ago. America keeps cutting its space research assignments every year. There are no serious assignments in the world for the invention of either the Supreme Mind or AI. Yet they are most important and perspective problems of the humanity. The computers that we have at present are used for modeling nuclear weapons and sometimes, weather. Furthermore, mankind does not spend much time thinking over the reason and goal of its existence. People spend a lot of their efforts and funds for solving local, temporal problems. Huge money and efforts are spent on conflicts and wars.

A certain hope has appeared recently. As experience shows, unmanned planes are a lot cheaper than piloted warplanes. More importantly, unmanned plane crashes do not cause harsh public reactions in civilized countries when pilots' or soldiers' deaths occur. The Americans design such planes successfully, but the planes are controlled by an operator within the United States. It has been proved that this remote control is not good for unmanned planes. The United States missed Bin Laden and Omar in Afghanistan several times, and two Iraqi pursuit planes downed an unmanned Predator in the Iraqi airspace. An unmanned plane can become something valuable

Figure 10.4 The future of pop music, this humanoid robot demonstrates its vocal ability.

indeed, if it has an AI, if it is capable of recognizing and destroying targets itself. The Pentagon has assigned certain money for the research of this issue. It is a difficult goal to pursue (to create the mind of a pilot), but it is a very prospective one. In this case, there would be no need to eliminate the young part of a country's population, if robots could conduct warfare.

The author suggested the hierarchical structure of AI, on the grounds of which the real brain probably works. Let us imagine a state with a dictator at the head. A dictator would never be able to find efficient solutions for external and internal state problems. A dictator has ministries, which are then divided into divisions and departments. This forms a pyramid, in which all departments have their own databases, as well as access to the common base. All divisions are busy with their particular problems, in accordance with the dictator's ideology. A dictator only sets problems up, while the appropriate divisions suggest solutions. For example, a man decides to cross a road with a heavy traffic. He looks at the road, while appropriate parts of his brain automatically receive the information about the width of the road, the distance to nearest cars, their speed, and so on. The brain automatically makes the necessary calculations, which eventually lead to the final decision: when is it safe to cross the road. All kinds of enlightenment in the solution of a problem are simply considered to be the "help from above." However, this is nothing but the joint work of that pyramid. Human beings do not even know that such a pyramid exists. The pyramid's decisions are based on the knowledge of a certain individual. If an individual knows absolutely nothing about the quantum theory, he will never solve any of its problems. Therefore, an AI of a high level cannot be realized with a personal computer that has only one chip and a successive work order.

The author's scheme stipulates the distribution of functions between parallel chips. The top one of them is offered to deal only with solution variants, their estimation, and choice.

Every human being wants to extend his or her life. This can be seen from everyone's wish to have children, or to do something outstanding. It is simply enough to avoid danger sometimes. Even suicidal terrorists believe that they will go to heaven when they kill themselves. The most important problem that the humanity has is the problem of immortality. Let us hope that it will be solved in the future.

10.7 Summary

1. The chapter discusses the problem of science, soul, paradise, and AI. It is shown that the soul is the only knowledge in our brains. To save the soul is to save this knowledge.

11 Real Breakthrough to Immortality

11.1 Brief Description of Previous Works by the Author

In a series of earlier chapters (see references at the end of the book), the author shows that the purpose of nature is to create Super Intelligence (SI). With its ability to understand the universe, advanced entities with SI power will be able to survive major cataclysms. There is the Law of Increasing Complexity (in opposition to the Entropy Law, increasing chaos). This law created biological intelligence (people). Humans have since become a sovereign entity on the Earth and in nature, above all other creatures.

However, humans are just as mortal as any other biological creature. The human brain and body include albumen, molecules containing tens of thousands of atoms united by weak molecular connections. A change of only a few degrees in temperature results in death. The human biological brain and body require food, water, oxygen, dwelling, good temperature, and a good environment in order to survive. These conditions are absent on most other planets. This makes it difficult for humans to explore space or settle on other planets. Humanity loses valuable information (human experience) with old age and death, and humans invest considerable time and money toward raising and teaching children.

11.2 Electronic Immortality, Advantages of Electronic Existence

In earlier works, the author has shown that the problem of immortality can be solved only by changing the biological human into an artificial form. Such an immortal person made of chips and super-solid material (the E-man, as was called in earlier chapters) will have incredible advantages compared to conventional people. His brain will work from radioisotopic batteries (which will work for decades) and muscles that will work on small nuclear engines. He will change his face and figure. He will have super-human strength and communicate easily over long distances to gain vast amounts of knowledge in seconds (by rewriting his brain).

As was written in the science fiction book *The Price of Immortality* by Igor Getmansky (Moscow, Publish House ECSMO, 2003, in Russian), an artificial person will have all of these super-human abilities.

Universe, Human Immortality and Future Human Evaluation. DOI: 10.1016/B978-0-12-415801-6.00011-6

11.3   What Are Men and Intelligent Beings?

All intelligent creatures have two main components: (1) **Information** about their environment, about their experience of interacting with nature, people, society (soul); and (2) a **capsule** (shell), where this information is located (biological brain, body). The capsule supports existence and stores information and programs for all of its operations. The capsule also allows the creature to acquire different sensory information (eyes, ear, nose, tongue, and touch) and to move it to different locations in order to interact with the environment.

The main component of an intelligent being is information (soul). The experiences and knowledge accumulated in the soul allows the entity to interact more efficiently in nature in order to survive. If the being has more information and better operational programs (ability to find good solutions), then it is more likely to thrive.

For an intelligent being to save its soul, it must solve the problem of individual immortality. Currently, man creates a soul for himself by acquiring knowledge from parents, educational systems, employment, and life experiences. When he dies, most knowledge is lost except for a very small part which is left through works, children, and apprentices. Billions of people have lived on Earth; however, we know comparatively little about ancient history. Only after the invention of written language did people have the capacity to easily save knowledge and pass it on to the next generation.

As discussed earlier, the biological storage (human brain) of our soul (information) is unreliable. The brain is difficult to maintain and requires food, lodging, clothes, a good environment, and education, etc. To support the brain and body, humans spend about 99% of their time and energy, and eventually what knowledge is gained is taken to the grave at death.

There is only one solution to this problem: to rewrite all of the brain information (our soul) in more strongly based storage. We must also give the soul the ability to acquire and manipulate information from the world. This means we must give sensors to the soul, so it may have communication and contact with people and other intelligent beings. We must give the soul a mobile system (e.g., legs), systems for working (hands), etc., thus giving the soul a new body in which to live.

The reader may ask, these ideas seem interesting, but how does one rewrite a human soul to live within a new carrier, for example, in electronic chips?

11.4   The Main Problem with Electronic Immortality: Rewriting Brain Information (Soul) to Electronic Chips Is Impossible to Do with Current Technology

At present, scientists are working to solve this problem. They know that the brain has about 15 billion neurons, and every neuron has about 10 connections to neighboring neurons. Neurons gain signals from neighboring neurons, produce signals, and then send these signals to others neurons. As a result, humans are able to think and find solutions. On the basis of this way of thinking, humans can come to solutions without exact data. (Concepts of the brain were described in the author's previous papers. For example, see "Locate God in Computer-Internet Networks" or "Science, Soul, Heaven, and Supreme Mind." See also the author's papers on the Internet and references at the end of this book.)

Scientists are learning how to take individual neurons on microelectrodes and record their impulses. The ideas of scientists are very simple: study how single neurons and small neuronal networks work and then model them with computers. They hypothesize that if we can model 15 billion neurons in a computer, they will learn how the brain works, and then they will have AI equaling the human brain.

In the author's previous works [1]–[4], this is shown as a dead-end direction for human immortality. It is true that we will create an AI that will be more powerful than the human mind. However, it will be his AI, and a new entity altogether. Our purpose focuses on preserving the concrete person now (or more exactly, his soul) in a new body in order to achieve immortality.

Why is it impossible to directly write the information of the human brain onto a chip? Because the human brain is constantly changing and neurons permanently change their states. Imagine you want to record the state of a working computer chip. The chip has millions of logical elements that change their state millions of times per second. It is obvious that if you write the current state of the chip in series (one after another), it is impossible to instantly write all states of the chip's elements. To instantly write all neurons, one would need to insert a microelectrode into every neuron, and this would destroy the human brain before the writing was complete.

In the article "Science, Soul, Parade, and Supreme Mind," the author offered another method for the solution of the main problem of immortality (MPI).

11.5 Modeling of the Soul for a Concrete Person

As previously stated, straight rewriting of a human mind (human soul) to chips is very complex. Straight rewriting will not be possible in the near future. All scientific works studying the work of human brains at the present time are useless for the MPI. They are also unworkable for the problem of AI in the near term, because the brain solves problems by way of general estimations. AI solves problems based on more exact computation and logical data.

To solve the MPI, the author offers a method of "modeling the soul" of a concrete person. This method does **not require interventions into the brain** of a given person. This method may be applied immediately at the present time. But an accurate modeling is needed, depending on the modeling period.

Before describing this method, let us analyze the human soul and what components are important for each person and his environment. All information in the human brain (soul) may be separated in two unequal groups: (1) The **memory** (permanent knowledge) about the person's life (all that has been seen, heard, made, felt, people met, his (her) behaviors, opinions, wishes, dreams, programs of activity, etc.), environment; and (2) **methods** of processing this information, i.e., producing new solutions and new behaviors based on this knowledge.

The first part (knowledge) is very large. It fills most of the memory and remains relatively constant (you remember your life, your history, and you can only fill it by what was in the past). The second part (methods for deciding, producing solutions based on your knowledge) is relatively small and constantly changing because of new information, facts, and life experiences.

However, the most important part of a human soul can be written without any problem now. Industry is producing cheap micro-video recorders as small as a penny, microphones the size of grain, and microsensors for vital signs (breathing, palpitation, blood pressure, skin resistance, perspiration, movement of body parts, etc.). These measurements allow for easy recording of not only the physical state, but of the moral state (joy, pleasure, grief, trouble, anxiety, nervousness, etc.). For example, lie detectors are able to define not only the state of a man but also the truth of his words. Now we can measure and record brain commands, and we can produce small cards with 4 gigabytes of memory (Figure 11.1).

It would be easy to attach a video recorder and microphone to a man's forehead and then attach sensors to the body and record all that he sees, hears, speaks, feels, his reactions, and his activity. We could then rewrite this information into a personal hard drive (long-term memory of high-capacity storage) at the end of each day. As a result, there is a record of the most important part our soul: a history of life, feelings, environment, behaviors, and actions. This would be more detailed than what is

Figure 11.1 Typical devices for writing information in the human soul. At present, a group of enthusiasts designs modern devices for permanent recording of information, environment, and the human state (for further information, write to abolonkin@juno.com).

captured by the real man, because humans forget many facts, feelings, emotions, and personal interactions. The electronic memory would not forget anything in the past. It would not forget any person or what they were doing.

But what about the second, smaller part of the human soul—the part that produces solutions based on personal knowledge? Perhaps the meticulous reader wants to know. This could be restored by using past information from the real man in similar situations. Moreover, an electronic man could analyze more factors and data in order to throw out and exclude actions and emotions that happened under bad conditions. The electronic man (named E-being in previous works) would have a gigantic knowledge base and could in a matter of seconds (directly to his brain) produce the right answer, much faster than his biological prototype. That means he would not have the need for the second smaller part of memory.

Considering the environment and friends, the following is an important part of a man's soul: his relationship with parents, children, family, friends, known people, partners, and enemies. This part of his soul will be preserved more completely than even his prototype. Temporary factors will not influence his relationship with his enemy and friends as would happen with his former prototype.

There is one problem which may be troubling for some: if we were to record every part of a person's life, how do we keep intimate moments a secret? There are (will be) ways to protect private information which could be adapted from current usage, e.g., the use of a password (known only by you). Also there may be some moments when you choose not to record information or decide to delete the information from memory.

The offered system may become an excellent tool for defense against lies and false accusations. You may give the password in one given moment of your life, which proves your alibi or absence from the accusations.

Some people want to have better memory. Video makes up 95% of storage capacity, sound 4%, and the rest 1%. In normal situations, video can record only separate pictures and sound only when it appears. This type of recording practice decreases the necessary memory by tens of times. But every 1.5–2 years, chip storage capacity doubles. There are systems which will compress the information and then one may select to record the most important information (as is done in the human brain). During your life, the possibility of recording all information will be available for all people. This type of recording apparatus will be widely available and inexpensive. It is possible now. The most advanced video recorder or DVD writes more information than a CD.

This solution (recording of human souls) is possible and must be solved quickly. By mass production, the apparatus will become inexpensive. The price will drop to about $300–1000. If we work quickly, we can begin recording and then more fully save our souls. The best solution is to begin recording children when they become aware of "I." But middle-aged and older people should not delay. Unrecorded life periods may be restored by pictures, memories, notes, diaries, and documents. Soul recovery will only be partial but it is better than nothing.

These records will also be useful in your daily life. You can restore recorded parts of your life, images of people, relatives, and then analyze and examine your actions for improvement.

11.6   Disadvantages of Biological Men and Biological Society

People understand Darwin's law "the survival of the fittest." For a single person, this law is the struggle for his/her personal existence (life, well-being, satisfaction of requirements, pride, etc.). In a completely biological world built on Darwinian law, the strongest survives and reaches his goal. Though they may be intelligent, humans are members of the animal world. They operate as any other animal in accordance with animal instincts of self-preservation. If one is poor, at first he struggles for food (currently half of world's population is starving), dwelling, and better living conditions. When one reaches material well-being, he may struggle for money, job promotion, reputation, renown, power, attractive women or men, and so on. Most people consider their activities (including official work) in only one way: what will I receive from it? Only a small number of people are concerned with the idea of sacrificing themselves to the well-being of society at large (seldom giving up their lives).

As a result, we see human history as a continuation of wars, dictatorships, and repression of people by power. Dictators kill all dissidents and opponents. Most people try to discriminate against opponents and play dirty against their enemy. There are murders, rapes, violence, robbery, underhanded actions, fraud, and lying at all levels of society, especially in lesser-developed countries. Each person only cares for himself and his family and does not care how his actions affect other people or society.

Democratic countries try to cultivate a more civilized society. They create laws, courts, and have police. Dictator regimes, on the other hand, make only the law they want. There are thousands of examples to verify this concept. But hundreds of millions of people are killed by war, aggressive campaigns, repressions, genocides, and thousands of criminals in the everyday world are a good illustration of this.

The human brain allows us to reach great success in science and technology. However, as a biological heritage, struggling for individual existence in a bloody, dangerous world, humans spend much of their resources on mutual extermination of intelligent beings. Moreover, humans have created ever powerful weapons (e.g., nuclear and hydrogen bombs) that could wipe out humanity. In time, existence may depend on the volition of one man, perhaps the dictator of a nuclear state.

The second significant drawback to the biological body is that it spends 99.99% of its effort and resources simply to support existence, such as food, lodgings, clothing, sex, entertainment, relaxation, environment, and ecological compatibility. Only a very small part is used for scientific development and new ideas and technology. The reader may see something wrong here.

States use a percentage of their revenue for research into science and technology. This percent is used not for new ideas, but is used to commercialize modern processes. All research is included in the state budget under the name "Science and New Technology." But much of this research has little relation to real new scientific progress. Even in the United States, states spend only a small part of the assigned money on new science because state officers do not understand the research. People, organizations, and companies fight for a piece of the pie. Geniuses are rare and usually do not have the capacity to move forward because they must promote and pay for new ideas from their own packets.

Yet, science and technology has seen success. Most advancement (90%) was made recently in the twentieth century when governments started to finance a few scientific projects (compared with the millions of years of human existence). However, our current knowledge and new technologies are far from what we will eventually have. The first government of an industrialized country to understand and realize the leading role of new science and innovation will become powerful.

11.7 Electronic Society

The electronic society will be a society of clever electronic beings (or E-beings). Most of the reasons and stimulus which incite men to crime will be absent in E-beings. E-beings will not need food, shelter, sex, money, or ecology, which are the main factors in crime. E-people will not have intense infatuations or be distracted by behaviors, because they will have vast knowledge about the open electronic society. Their main work will be in science, innovations, and technologies. They will save their mental capacity for the production of chips and bodies, scientific devices, experimental equipment, space ships and space station, etc. They will need a number of robots, which do not need a big brain. It is likely they will award these robots better minds and memory. It is also likely that the E-man will unite in a common distributed hyper-brain, which will become a sovereign of the Universe (God).

Nature is infinite and the development of a Super Brain (God) will not be limited. On the other hand, biological people will have limited mental capabilities. It will be difficult for them to imagine and predict the development and activity of Super beings, which we will generate.

Many, especially religious people, object because they say electronic beings will not have human senses such as love, sympathy, kindness, humanism, altruism, and the capacity to make mistakes. E-beings are not people. Look back at human history. Human history shows that kindness plays a very small role in human life. All human history is the history of human vices and human blood: struggle for power, authority, impact, money, riches, territory, and states. All human history is filled with fraud, underhanded actions, and trickery. Ordinary people were only playthings, flocks of sheep for the tyrants and dictators.

Some people object that with an electronic face, humans will lose the joy of sex, alcohol, narcotics, appreciation of art, beauty, nature, etc. The author answers this question in his article "Science, Soul, Heaven, and Supreme Mind" (http://Bolonkin.narod.ru). The brief answer is that electronic humans will enjoy all this in a virtual world or virtual paradise. Time will run millions of times faster in the virtual world. E-man will spend a few seconds of real time and live millions of years in the paradise. He will enjoy any delight

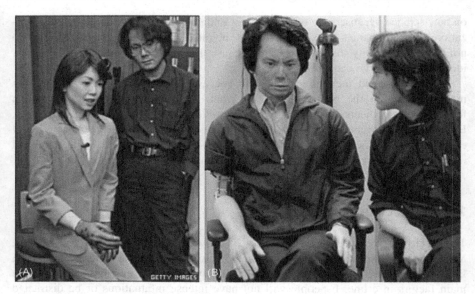

Figure 11.2 (A) "Actroid ReplieeQ1-expo" at Expo 2005 in Aichi, (B) with cocreator Hiroshi Ishiguro (2000).

imaginable, including sex with any beautiful women (or handsome men), and feel the emotions of any commander, leader, criminal, or even a dog.

11.8 The Lot (Fortune) of Humanity

Biological humanity will be gradually transformed to electronic beings. Old people, when their biological bodies cannot support their brains, will continue their existence in electronic bodies after death. They will become young, handsome, and robust. Fertility in biological men will decrease. Birthrates are lower than death rates in many civilized countries now (e.g., France) (Figure 11.2).

Population growth is mainly supported by emigration from lesser-developed countries. When education levels increase, birthrates will fall.

For a time, biological and electronic people will exist together. However, the distance between their capabilities will increase very quickly. Electronic people will reproduce (multiply) by coping, learning instantly, and will not need food or dwellings. They will work full days in any condition such as in space or on the ocean floor.

They will gain new knowledge in a short time. They will pass this knowledge on to others who do not have enough time. The distance between biological and artificial intellects will reach a wide margin so that biological people will not understand anything about new science just as monkeys do not understand multiplication now, even after much explanation (Figures 11.3–11.6).

Figure 11.3 Artificial girls.

Figure 11.4 ASIMO is a humanoid robot created by Honda. Standing at 130 cm and weighing 54 kg, the robot resembles a small astronaut wearing a backpack and can walk on 2 ft in a manner resembling human locomotion at up to 6 km/h. ASIMO was created at Honda's Research & Development Wako Fundamental Technical Research Center in Japan (2003).

Figure 11.5 Artificial girls.
Source: http://world.honda.com/news/2005/c051213_8.html.

It is obvious, and clever people will see that there will be a huge difference between the mental abilities of biological and electronic entities. They will try to transfer into electronic forms and the ratio between biological and electronic entities will quickly change in electronic favor. A small number of outliers will continue to live in their biological bodies in special enclaves. They will not have industrial power or higher education and will begin to degrade.

Naysayers may promote laws against transferring into an electronic man (as cloning is forbidden now in some states). However, who would renounce immortality for themselves, especially while they are young and healthy? One may denounce immorality as blasphemy, but when you (or your parents, wife, husband, children) die, or especially if you are near death yourself, one comes to understand that life is extremely important. The possibility to live forever, to gain knowledge that improves life, will also allow one to become a sovereign force in the universe.

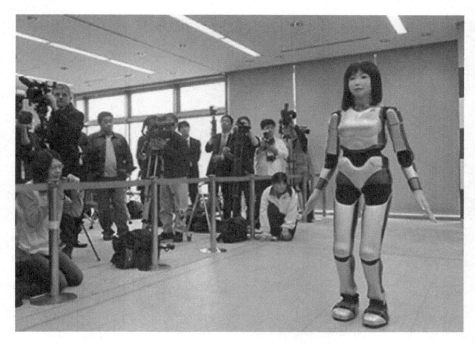

Figure 11.6 A "cybernetic human" HRP-4C, designed to look like an average Japanese woman, walks in front of journalists during a demonstration in Tsukuba, near Tokyo, Monday, March 16, 2009. The humanoid robot having a female face and black hair trimmed down to 43 kilograms (95 pounds) to make a debut at a fashion show later in the month. http://www.washingtontimes.com/news/2009/mar/16/female-robot-hit-japan-catwalk/

11.9 Summary

1. The author offers a new method for rewriting the human brain on electronic chips. This method allows for the modeling of a human soul in order to achieve immortality. This method does not damage the brain but works to extend and enhance it.

Figure 11.[illegible]

11.9 Summary

1. [illegible]

12 The Natural Purpose of Humankind Is to Become God

12.1 Introduction

In a series of articles (see publication list at the end of the book [1]–[7], [13]–[26]), the author examined some questions about human immortality and a totally electronic civilization. In those chapters, the author briefly mentioned the law of increasing complexity, self-copying systems, and the purpose of human existence. In this chapter, the theme is presented in more detail, and thus developed and improved.

12.2 The Law of Purpose

The real purpose of aggregations of living matter, for a long time, has fascinated philosophers. Various purposes seem obvious to the overwhelming majority of people, and they are seen as constant guides in daily life: the struggle for well-being (riches) by the individual and the family, the search for mates, for sexual or other pleasures, for glory, authority, etc. These aims may be thought of as local or individual purposes. Only insignificant numbers of people pursue aims that reach higher than those of their particular group or specific community. But these too can be reduced to personal purposes, to a mere desire for popularity, glory, or authority. Charles Robert Darwin (1809–1882) summed up these purposes with the general term *struggle for survival*, understanding it first of all as the struggle for existence of living creatures against other types of life forms. Any human has personal, local, regional, spatial, and temporal purposes that can vary depending on geographical location, on the civilization's developmental period, and on immediate circumstances. For example, if a person is hungry, the first purpose will be food. If the person is fed, the next purpose may become the pursuit of pleasure, or more a distant purpose—riches, glory, or authority.

In this chapter, we will examine only the global purposes of intelligent life or the more overall aims of reasonable human beings, alternating between the concept of biological reason and reason that is artificial, electronic, and self-developing.

People easily understand individual, personal, and local purposes; group purposes are more poorly understood, and the purpose of a global human society and geopolitical state is even less understood. People seldom think of the purposes of humankind,

Universe, Human Immortality and Future Human Evaluation. DOI: 10.1016/B978-0-12-415801-6.00012-8

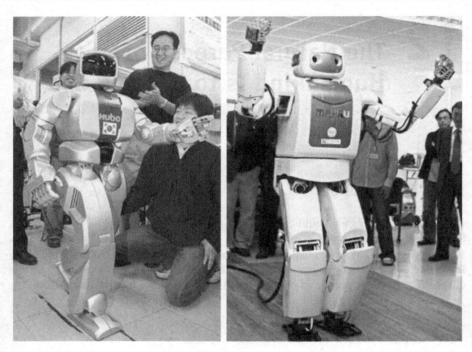

Figure 12.1 The creation of humanoid robots marks a brilliant advance in Korean robotics.

or life as a whole, in the course of their daily activities. If they can achieve their primary (individual) purposes, they may attempt to influence the secondary (societal) aims, but the tertiary purposes (of humankind in general and all life in Earth's biosphere) are seen as beyond their sphere of influence. About these, for the most part, they are ignorant (Figures 12.1 and 12.2).

The third purpose is defined by Nature. People can study and operate only in accordance with natural laws. As will be shown, nonobservance of such laws, and furthermore counterreaction by Nature, can result only in slavery or worse—the disappearance of the given kind of beings or reason itself.

Thus, it is possible to formulate the following law:

Any kind of life or reason has a global purpose determined by Nature.

What this purpose is will be considered later in the chapter.

12.3 What Is Alive and What Is Imbued with Reason (Intellect)?

For further consideration, we should specify the concepts "alive" (life) and "reason" (intellect, mind). Under "alive" we understand an essence (or community) capable of reproduction. Bacteria, plants, and aggregated discrete packets of living matter

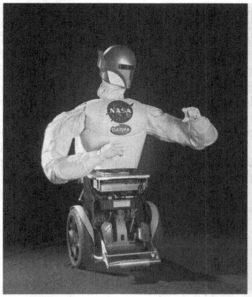

Figure 12.2 Robonaut B in December 2003.

(in the usual sense) or biological essences, including animals and humankind, fall under this definition. Artificial electronic essences (E-beings) may also be living creatures when we have taught them to reproduce themselves, that is, when they can live and develop (as a society or a reasonable essence) without our further participation.

We understand essences as independent, reasonable beings capable of building theoretical models of their environment, of predicting their long-term behavior, and of transforming it in their own interests. These three criteria, based on Earth's biological essences, are answered only by humans. AI of the kind that exists today can meet the first two criteria, but such intelligence is not yet engaged in questions of engineering, supply, or self-manufacture, dooming it to the role of servant to the biological person.

12.4 Biological Intellect as the First Step Toward the Purpose

In Ref. [1], the law of increasing complexity of self-copying systems was formulated. All life on Earth today confirms the existence of this law. After the great invention of Nature—self-duplication—life appeared and was rather quickly improved, eliminating everything that was not adapted to external conditions (the generalized law of Darwin's "struggle for survival"). Microbes, then plants, then animals appeared. Eventually, Nature made the next great invention—humanity (intellect).

Since its inception, humanity has actively altered the natural world according to the needs generated by its own existence. Unfortunately, humanity has also inherited a heavy burden (in a deterministic sense)—a struggle for the best possible personal existence—a monstrous set of individual purposes, emotions, and passions that sharply impede progress toward a harmonious global society.

In previous articles, it was shown conclusively what must be overcome. This contradiction demands transition to a higher level of reason (intellect)—to an electronic society that will be relieved of many defects of a human society (money-making, individual greed, sexual instincts, aspiration to authority, racial and religious conflicts, boredom, etc.), a society that will not require food, shelter, air, an unpolluted environment, or health services, that will not spend decades in the cultivation of land or the educating of its posterity or all the huge effort this implies (roughly 99.9% of its time and energy). An electronic society will be able to live on the majority of planets without air, water, or energy from suns.

12.5   What Is God?

Who God is, whether he represents himself in some sense, where he dwells, clearly no one knows. Believers and attendants of religious cults can only say that God is an omnipotent, reasonable being. Actions attributed earlier to God, for example, thunder, lightning, creation of the world and its inhabitants, have been established by science as simply natural phenomena that submit to certain physical laws; some of them, for example, thunder and lightning, may be reproduced artificially.

In the consciousness of many people, God is associated with a certain reasonable omnipotent being that can affect everyone. But is such a being possible? The elementary question, "Can God create a stone that he cannot lift?" at once stumps theologians of all organized religions. If he cannot lift it, is he still an omnipotent being? If he cannot create such a stone, the coercive force of God is relative. He seems to us all-powerful only in the deciding of such problems, which to us are not compelling—and are probably nonexistent.

The second point, beyond which nobody reflects but is part of everyone's consciousness, is that: above God there is no authority to which humans submit. Otherwise, for God the omnipotent creator there is a stronger essence to which he is compelled to submit and whose will he must obey. Below God are subordinates—for example, angels, who are said to have a certain power, but more or less carry out the will of God.

So, we may understand God as *the strongest reasonable essence, whose actions are limited only by the laws of Nature.*

For the sake of expediency, Nature here may be seen as analogous to the English Parliament, the American Congress, or the Russian Duma. It establishes physical laws that are identical everywhere and cannot be broken even by God acting as President of a given period and region, which is possibly also all the known universe.

Local Gods are possible (probable) while there are no contacts and the God does not know about other Gods in other parts of the universe. But, as soon as facts about them become known, the main God becomes stronger (the God of a region with higher scientific and technical power). In this sense, the God is always one, and all others (at best) can be only angels.

12.6   God as the Purpose Given to Humankind by Nature

But if we agree that God is the strongest essence (in terms of opportunities in reorganizing an environment), from here at once it generally follows that humanity (or rather, human society as a whole) is a God in this solar system. Nobody doubts that humans are the most reasonable and powerful essence on Earth. Humanity is reasonable because it has learned much about the devices of the world, has constructed valid theoretical models of its environment beginning with a microcosm and ending with a model of the universe. Humanity actively uses these models and theories for altering the Earth, and for flights into surrounding space. People have created powerful industry, huge geographical regions planted with the crops necessary to feed and clothe themselves, and they manage vast herds of cattle and other animals. In relation to the living terrestrial world, the human is the God, in a position to liquidate (or to make happy) any other organism or even a whole species. As an example, with a single kick a person can destroy an anthill that required years to build. (A religious ant, seeing only a distance of 1 cm, might perceive it as an act of nature or a divine punishment.) But the person (i.e., humankind) is God in this solar system; we know clearly that no other being in our solar system matches humanity in terms of intellectual development. There may not even be microbes on those other planets. About other reasonable essences in our Milky Way galaxy or the universe, we know absolutely nothing.

But some may object that many people are unfortunate and lack the basic necessities. But do you suppose that God or his angels are all happy? You think so because they can solve your problems at one stroke. But they have problems in which they too are engaged that make them happy or unfortunate.

Nature has made humankind the strongest biological essence on Earth and in this solar system, that is, the local God. And if humanity wants, if we are not enthralled by other Gods (as all alive on Earth have been), we have a unique opportunity—*to become the God of our Milky Way galaxy, and later to be the God of the universe.* This global and unique purpose of humanity is given by Nature. And the sooner humanity realizes this aim, and aspires to it, then the better humanity's chance of avoiding slavery by beings of higher reason (aliens?) and of not ending up in a category of the lowest reasonable essences.

12.7   An Electronic Civilization as the Second Step of Reason (Humanity)

For creation of the first biological reason, Nature used a single method, known in science as "trial and error." This testing method is markedly inefficient. Nature has spent hundreds of millions of years doing billions and billions experiments. As a matter of fact, each of billions of possible connections of atoms and molecules was a trial or experiment. The first revolutionary break was made when self-reproducing organisms (viruses, bacteria) began to combine into colonies and later into plants and animals.

The second break occurred when Reason carried out a purposeful selection, accelerating the promotion of Purpose by millions of times.

However, the biological and reasoning civilization spends only an insignificant part of its resources in the movement toward Purpose. As biological essence, the person requires food, shelter, heating and cooling, rest, entertainment, sex, and sleep. A person is trained extremely slowly, often learning only by oversights, mistakes, etc. (Figure 12.3).

Humankind spends 99.9% of its energy and resources on maintaining its existence. Only 0.1% of resources go toward development of new engineering and new technologies. But at the heart of human scientific and technical progress, a new

Figure 12.3 Sheikh Mohamed bin Zayed, Robot REEM-B, and Prof. Noel Sharkey.

revolutionary break is being planned that will speed up the scientific–technical progress by a thousand times and will subsequently allow us to proceed to a new kind of Reason—an electronic civilization. More detail about this is provided in the author's and others papers [1–30]. Journalists and visionaries have a fairly impaired image of electronic Reason—representing it as a kind of stupid clumsy robot incapable of competing with the "clever" person and at best being his or her servant. And while it lacks the complex of abilities that characterize the person, in simple problems (tasks) amenable to algorithmization, the computer works more speedily and more consistently than the best-trained person. While the computer does not have enough comprehension to have its own "I am," its own interests, it may yet be given sensors for studying an external world and "hands" for its own reproduction and perfection [4]. All this is acquirable (i.e., coming with time's passage). According to Gordon E. Moore's law in 1965, every 1.5–2 years, the speed and memory of computers is doubled. Capacity (operation speed) of supercomputers has already passed 100 teraflops, and a capacity of more than 1000 teraflops is projected (as of 2005). So E-essences (see Refs. [1–27]) are not only achieving a human level but also will exceed the human level.

12.8 Electronic Immortality as a Way of Transition to an Electronic Civilization

The overwhelming majority of people intuitively feel and see in AI the enemy, which can surpass and supplant the person from his command positions in the local world (from the status of local God). They feel this will subordinate humanity, at best using us as we now use cows, sheep, hens, and other domesticated animals, which are lower on the scale of our own intellectual development. Meanwhile philosophers, journalists, and other writers assure us with fairy tales—that computers are machines that work only under the goad of programming and basically cannot be cleverer than us, as though the brain of a person is filled with anything but programs of training, knowledge, life experience. The person in all typical situations uses knowledge (programs) and acts (reacts) typically. Emotions are only a response to actions and situations.

But the brain of humans, practically, has not changed (in terms of memory size and speed) for at least the last thousand years, while the abilities of AI are doubled every 1.5–2 years. The winner in such a competition is obvious. A person's fears about his or her destiny as biological essence are well founded. But to block development of AI, to halt the movement toward the Great Purpose of Nature, is a refusal to be the God of the universe, to doom ourselves to enslavement or even destruction by other, more developed alien electronic civilizations—which also are not an exit. This is death and impasse.

In Refs. [3–7], the author offers a unique outcome preferable to such an impasse—the gradual transition of humankind into an electronic immortality. The person lives the usual biological life, the history of which is fully entered in chips, and at the end of life all his or her history is located in an electronic brain and he or she continues to live already in a new electronic form. In this form, the person does not require food, shelter, water, air, or sleep. He or she can travel in space outside the Earth's biosphere or at the bottom of an ocean without a survival suit, can be supported by nuclear batteries, change shape at will, undertake out-of-body travel on other planets (teleportation), and copy the contents of the brain (soul [3]) to the body rented there with the help of a laser beam. The human becomes immortal and cannot be destroyed by any weapon because he or she can store all contents of the brain (soul) separately and may be restored after full destruction.

In Ref. [3], the main problem is solved—how to copy the basic maintenance (contents) of a person's brain into chips using already-existing engineering and without the intervening activity of a brain.

Only an idiot will refuse immortality! Besides the second obstacle, fear too is removed, that electronic reason will enslave biological mankind. E-essences will remember their origin and hardly want to enslave or destroy their parents and relatives. Imagine what might concern you if you recall that in the past, while still a monkey, you skipped and jumped on trees. Most likely the birthrate of people will fall or the biological civilization will be limited—but it will be gradually transformed into an electronic civilization.

12.9   What Can We Expect from Other, Alien Civilizations?

Many people put big hopes in searching for and seeking the help of other, more advanced civilizations. People think about biological civilizations automatically, and usually assume their shape to be close to the shape of people. Well, perhaps the nose, chin, eyes, or ears are unusual! Somehow it is assumed that because these aliens are more developed, they are also more humane and will immediately share their knowledge and help us.

The author wants to show that in scientific and technical progress, humankind can count only on itself. Technically backward civilizations and even civilizations equal to ours can give us nothing. More advanced civilizations will be only a sign for us that our civilization has lost the space race to the Supreme Mind (Reason), and we are now awaiting enslavement and ultimate disappearance. Imagine a scenario in which you meet a certain primitive human tribe living in caves, and they ask you to share your knowledge. You try to explain a nuclear reactor, or an airplane, or a computer; you try to tell them how it is possible to communicate by radio or TV with places thousands of kilometers distant. You will not be understood, and even if they believe such things are possible, all this knowledge is useless for them, because using them entails having a big state with a population in the hundreds of millions of active persons, with a coexisting power industry (i.e., it is necessary *to have* this knowledge *already*). It is necessary to train hundreds of thousands of scientists, engineers, technicians, material workers, all of which requires hundreds of years of learning and huge physical and material means. In the best case, you can train them in producing bows and spears with bone tips. But most likely, they already have guessed how to do this—and they can do it better without your help.

You will understand that this primitive tribe living is behind us technologically by some thousands of years. But go back 200–300 years, back when electricity was basically unknown (the first galvanic cell was not invented until 1799 by A.G.A.A. Volta (1745–1827)). And the newcomer from space starts explaining about transfer of energy on wires, or communication and sending of images by means of electromagnetic waves, or the working of an electric motor. And you have no concept at all; there is no electricity, and there is no electric–radio–television industry. How could you possibly understand and/or use this knowledge? You would need 200–300 years; you would need a lot of money to create the scientific-engineering staff and to construct the appropriate industries. For this time gap of knowledge and industry, the aliens will be so much more advanced than you that you cannot compete with them.

And why would the space newcomers share their technologies? Imagine that astronauts have found monkeys, cows, or pigs on Mars. How will they train these creatures in human knowledge? So why do people not do this on the Earth? And if people plant and feed them, they do so only to use them as food, or to receive milk, wool, or eggs. All fauna of the Earth have lagged behind the development of humans and have become slaves. They exist only in the frameworks allocated to them by human beings and only in the interests of humankind. Moreover, people do not want to share the knowledge and high technologies even with other people and states on Earth. Numerous secrets, patents, and know-how serve the purpose of keeping these achievements in the technically advanced states. And this is understandable: if the advanced states keep secret their manufacture of explosives and weapons of mass destruction, terrorists with only bows and spears cannot render them much harm.

12.10 The Great Space Race

In scientific and technical progress, the person should hope *only* for himself. Moreover, humankind can succeed *only* in the one case that it is the most scientifically and technically advanced, with the most powerful industry in the universe. Whether it wants to or not, humankind (and then an electronic society) is compelled to participate in the great space race of knowledge and technologies.

Humankind has achieved a power of the local God in this solar system. Our main task and the purpose imposed on us by Nature is to become the God in our galaxy, and then in all the universe. This is our greatest happiness, that the space newcomers have not yet arrived. That is a sign that we are the most advanced, knowledgeable, and technically advanced even in our Milky Way galaxy. This race is infinite because knowledge is unlimited. It does not mean that humankind will preserve its biological environment (Earth's biosphere). First, humanity is transformed into an electronic society, then a process of growth of knowledge and technologies occurs— in (maybe) proton, quantum, or quark society, and so on indefinitely. Each step will be a jerk forward on the basis of new knowledge and technologies, and each step will accelerate scientific–technological progress by hundreds and thousands of times. Most likely, all of the advanced community will represent a certain distributed collective Reason built on a common base of knowledge. It is possible that this Reason (Supreme Mind) will achieve such power that it can create a new universe, and even operate the laws of such a universe.

It will hardly be the God of present human understanding, who is interested in each person separately and sponsors him or her. We, being Gods in a solar system, are not interested in the life of each ant or even a separate anthill. We solve the global (from the point of view of an ant) problems: to cut down woods, to plough up the ground, to plant gardens, to irrigate deserts—not beginning from the existence on this ground of numerous anthills. Religious ants will no doubt view these destructive events as acts of nature or divine punishment.

12.11   One Alternative—God, or Slavery and Destruction

All of this can shock people, especially believers. They shout: "Where is humanism, kindness, mutual aid, feelings, etc.?" To all these emotional concerns there is one possible answer: look at human history, at all these uncountable bloody wars, the struggle for authority and power, bloody dictatorships, money-making, deceit, murder, terrorism. Human society is not ideal. It is amazing that our society has made any scientific and technical progress at all.

But there can be another purpose for humankind. Certainly, everyone will speak of a purpose that is favorable to him or his estate: churchmen, that it is necessary to pray, to endow churches more so that the God will give all; communists, that it is necessary to work more and suffer need for the sake of a bright future and general happiness; party leaders, that it is necessary to vote for them so they can solve all our problems. As for the global purposes of humankind, it is still uncertain.

But an elementary analysis of the history of life and scientific and technical progress on Earth shows that humanity has only one alternative: to be the God of the universe, or come under the authority of a stronger (in the sense of alien knowledge) God who was born on another planet, in another civilization.

And the last option means only one thing—slavery, the loss of independent development (as all fauna of the Earth lost it after the surging ahead of humankind) and finally, destruction or extinction.

Whether we want it or not, we are participants in the Great Space Race to the God (or great and almighty Reason). Probably, we are ahead of all the nearest universe known to us (signs or attributes of other Reason are not yet revealed). And we should keep this leadership if we want to exist. After all is said and done, the Law of the Purpose can be formulated in a final way:

Any kind of life or reason has a global purpose determined by Nature. This purpose is the creation of a strong real God or Supreme Mind (Reason), which will rebuild the environmental validity under itself.

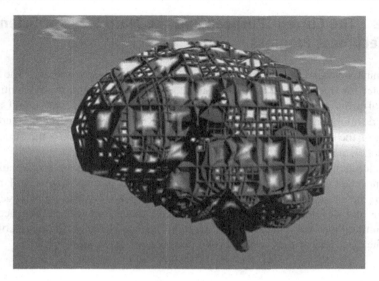

Figure 12.4 The possible form of the Superbrain.

The present organized religions, based on the notion that the God already exists and cares for people, dooms humankind to slavery, to the position of cattle on a farm for which the owner cares. Even if it is so, the God cares not because of altruism, kindness, or love of an animal, but in the way a farmer raises cattle for the sake of meat or milk, wool, hides, eggs, etc., and at the necessary moment sends the cattle to the slaughterhouse. Religion plays the important positive role in creating public morals (if it does not preach murder), but has a negative role in propagandizing—that the person is a slave of God (Figure 12.4).

Such a prospect, the suggestion and comprehension of slavery, is the worst thing possible to consider for the freedom-loving person.

The present war with terrorism is a war of the old world, which has begun to understand that its end will be fast, that it will soon be supplanted by the new scientific–technical world. Islamic terror is like a condemned man, the last and desperate means by which a retarded world tries to delay, stop, or divert scientific and technical progress.

More than a decade ago, the author wrote that AI sooner or later will far exceed human abilities, and he was derided. But already now half of the experts, setting artful questions to the computer, cannot tell whether an average person or the computer has answered them. The majority of them cannot determine with whom they talk on the Internet. And a gold medal and $100,000 (the Loebner prize) will be received by the creator (founder) of the first AI. Also, Japanese scientists have produced a humanoid robot: a beautiful woman who (while sitting) reproduces a woman's movements with perfect mimicry. And you start to understand why prescient Americans after the film *Artificial Intellect* by the Hollywood filmmaker Steven Spielberg (a film that mocks AI) have created a Society for the Protection of Robot Rights.

12.12   The Current Purpose and the Main Ways of Scientific and Technical Progress

What must we make for this purpose now? Thoughtful readers will ask. The answer is simple: for this purpose it is necessary to pay greater attention to scientific and technical progress, its mainstreams. From an overall objective, we should aim for greater development of computer engineering, knowledge of the structure and development of the universe, and the ongoing study of the microcosm—the structure of chemical elements, elementary particles, etc.

Development of computer technologies will allow us to make a qualitative and very important jump—having invested humankind in immortality, it will allow a speeding up of scientific and technical progress repeatedly. The knowledge of the universe and development of beyond-the-planet space engineering will accelerate the assimilation of the universe while the totalized knowledge of the microcosm will allow the discovery of new materials and powerful energy sources.

12.13   Summary

1. Precise comprehension of the role of humankind in Nature is extremely important for the natural purpose of its existence—to enable a correct choice in the general movement and direction of human society. God as an able essence is none other than higher Reason (Supreme Mind). It will be created by scientific and technical progress. Humankind has already become the God (the Supreme Mind, Reason) in our solar system, and Nature has given us an opportunity to participate in the Great Space Race of Reason to become the God of our Milky Way galaxy, and then possibly also the God of the universe in the form of Great Reason.

2. Humankind should soon realize the applicability of this concept—this great natural purpose, the opportunity given to him or her to do everything, to occupy a position of leadership, to not end up in slavery under higher civilizations, and to not disappear from the universe. In this race, humankind can hope only in itself. The only means of victory is our own scientific and technical progress and knowledge of the world environment around us.

13 Setting God in a Computer-Internet Net

13.1 Humankind as Logical Device

In previous chapters, the author has shown from computer sciences how the individual human is a biological logic device for the treatment of information. Our heads are a storage complex of information, practice, habits, views, political and religious creeds, memories, and programs saved and produced during our lifetimes. Each brain is a logical device for reprocessing this information. Our eyes, ears, skin, and other sense organs are sensors of information about our immediate surroundings. Our arms and legs are the executive organs of our brain's orders.

Let us resume: If we save all the information received and produced during a person's life, write it in a more stable medium (e.g., electronic chips), we save each person as an individual (a soul, if you like) for an infinite time. If we give such persons sensory equipment, then we allow free development as an individual person. If we add on executive organs, then the person has the capability of active influencing its surroundings (again, for an infinite time).

Computations and graphs have already been shown (see the author's works in Reference section).

Computer technology develops very quickly. The power and memory of computers will be equal to the power and memory of a human brain in 15–30 years. Later they will reach the power and memory of humanity as a whole and surpass it.

13.2 God as Superintellect

Religions appeared when humans acquired Reason. Uncomprehending the experience of natural phenomena, people explained it all as being the activity of some superbeing that controlled everything in the world, and they opposed it as an evil force. Humans needed a defense against these incomprehensible forces and searched for this defense in God. As mediators in communication with God and commentators on God's actions, priesthoods arose. Religions have had both a positive and negative role in human history.

Humanity, consciously or unconsciously, understood God as a reasonable, kind being with gigantic possibilities for helping in human activities.

Universe, Human Immortality and Future Human Evaluation. DOI: 10.1016/B978-0-12-415801-6.00013-X

Unfortunately, this kind, superclever superpower being is not very interested in our common problems. We must self-solve our everyday problems. He does not even want to sustain our faith. On the other hand, he is said to have announced in advance actions that would break the natural law (e.g., stopping the sun for a short time). Such moves apparently assure humanity about the existence of God and his omnipotent powers.

Most believer-scientists do not imagine God as an old man sitting on a cloud surrounded by angels. Rather, they understand God as a novice in reasoning who created the universe and gave us the capabilities of self-development and rebuilding the universe in accordance with our knowledge, ideas, and concepts.

13.3   Computers in the Modern World and Their Development

The first electronic computer was created in 1949. From this moment, humanity entered a new chapter of its history on Earth—an era of electronic logical operations, which is the basis of any intellect. During the next 60 years, these electronic logical devices would make gigantic developmental progress previously unimagined. The power and memory of computers have, since 1949, increased by tens of millions of times. The cost of one logical operation decreased some millions of times. The volume of a computer's processor decreased by thousands of times. And there is Moore's Law: every 1.5–2 years, the power and memory of computers double. Computers are used everywhere in economics, banks, machine-building and control, navigation, in the military, and in outer space.

The power of computers also grows quickly. In 1994, the author predicted that a supercomputer of 1 teraflop (10^{14} byte/s, equivalent to the human brain) would appear in 2000, but it was actually produced by Intel Corporation in December 1996.

In 1998, IBM received an order from the United States Department of Energy and produced the supercomputer "Pacific Blue" with the power of 4 teraflops. This is the equivalent of 15,000 personal computers (PCs), and its memory is 80,000 times more than any PC ever had. That memory is enough for storing the full text of all 17 million books collected in the world's biggest library—the United States Library of Congress. The computer uses 5800 processors.

In 2000, the United States Department of Energy ordered the new supercomputer "White Version," having the power of 10 teraflops and a cost of 85 million.

By the beginning of 2000, IBM announced funding of $100 million to produce the supercomputer "Blue Gin" at 1000 teraflops. They planned to finish in 2005. This supercomputer will have a height of 2 m and occupy an area of $4 \, m^2$. It will perform billions of operations per second and will be equivalent to 2 million modern PCs. That means it will be more powerful by several factors than all Russian computers of that time, and 1000 times more powerful than the famous "Deep Blue" (1 teraflop) that defeated world chess champion Garry Kasparov in 1997.

The computer contains 1 million microprocessors, every one of which performs 1 billion operations per second. In terms of programs, the given supercomputer is

Figure 13.1 (A) Today, Japanese robots already know how to talk and do basic housecleaning. (B) Toyota has unveiled a new "violin-playing robot"—the newest addition to its team of Toyota "partner robots" being developed to support everyday life.

equivalent to 1000 high-quality, highly educated specialists capable of illuminating any scientific area in a matter of seconds.

This supercomputer is used for modeling new nuclear weapon designs (real nuclear testing is now forbidden by an International Convention), for weather forecasting, for finding the structure of DNA, and for the synthesis of new medicines and materials. Since 1993, the fastest supercomputers have been ranked on the TOP500 list according to their LINPACK benchmark results. The list does not claim to be unbiased or definitive, but it is a widely cited current definition of the "fastest" supercomputer available at any given time. The K computer is ranked on the TOP500 list as the fastest supercomputer at 8.16 petaFLOPS. It consists of 68,544 SPARC64 VIIIfx CPUs, using the Tofu interconnect. It does not use any GPUs or other accelerators, and is one of the most energy-efficient systems on the list.

The **K computer**, named after the Japanese word "kei", which stands for 10 quadrillion, is a supercomputer being produced by Fujitsu at the RIKEN Advanced Institute for Computational Science campus in Kobe, Japan.

We have already mentioned hardware. Let us now consider the development of programs.

At the present time, programmers are having great success in the design of AI components that serve in the communication between man and machine. They have created programs that recognize speech, images, and voices; that produce scientific, technical, and economical computations; control bank transactions, industrial processes, and single aggregates; translate from one human language to another; and synthesize human speech. There are many computer game programs, some of them very complex. Self-improvement programs are also developing rapidly. All such programs can be the base for AI (Figure 13.1).

For example, there is a program called "BACON," in which measured data are input and then "BACON" finds the law they obey. When the observed positions of planets are input, the program gives the Keplerian and Newtonian laws. It took humanity thousands of years to make these discoveries. The computer spent only a few seconds.

In 15–20 years, PCs that reach the power of 1 teraflop (equal to the human brain) will be available for a current cost of $1000–1500.

13.4   Computers and the Human Brain

The human brain contains about 10 billion neurons connected one to another and working as logical and memory elements. The operation speed of these elements based in chemical reactions is small, about $1/100$ s, and the information transfer speed is low, about 30 m/s. That cannot compete with a computer chip that produces about 1 billion operations per second with an information transfer speed by electric current of about 300,000 km/s, roughly the speed of light.

However, human neurons collect in groups and hierarchical structures and work simultaneously like thousands of computers, operating the image data with control by a central processor. For example, if we want to get up out of a chair, we do not think about how many degrees we must incline our body, how to balance it, how to move our legs, what muscles we must exert, etc. We tell a subsystem of our brain that we want to stand up, and corresponding sets of neurons (chips) and gigantic numbers of subprograms guide the muscles, balance, coordinate the movement, check the positions, provide feedback, etc. Our brains, like any good computer, contain a huge number of subprograms for walking, running, working, swimming, recognizing speech, sounds, images, etc., and for producing ready opinions and knowledge—everything that we have studied, obtained, remembered, or created during our lives. All the standard solutions, flashes of inspirations, genius solutions, which come supposedly from "heaven" (or in our sleep), are the result of activity in subsections of our brain that were commanded (though we are not conscious of it) and put to work; and the work they do is more intense than the trouble a particular problem presents us.

If PCs continue to be built with the current architecture (all operations are performed only in the central processor), it will be difficult for PCs to compete with the human brain.

But modern supercomputers have thousands of conventional chips working parallel, with multilayered organization and governance. Unfortunately, they are very

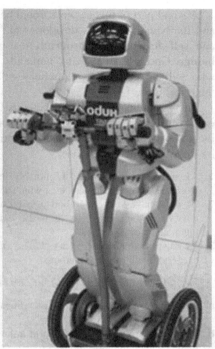

Figure 13.2 Humanoid robots.

expensive (tens of millions of US dollars) and they are produced not for the purposes of AI, but for the computations of nuclear weapons.

In addition, it is not yet clear how much power and memory are enough for a Superintelligence, and human beings have not yet created the program that will give birth to this Superintelligence.

13.5 Self-realization and Self-development Programs as a Germ of God

At the present time, humanity has the computer power, computer memory, and storage devices for human knowledge that are adequate for designing a Superintelligence, or in other words, a God. Today, there are about 200 million PCs in world, connected by orbiting satellites and the Internet in a united communications network. All current supercomputers are also connected to this net. The net is quickly growing and developing. Currently, about 90% of the time and memory of this net is unused.

A small part of this network could be allocated for research into Superintelligence and the beginning of its development. The problem is to design a **self-realization** and **self-development program** that will develop into Superintelligence, into a **benign God** who will thank humanity for its realization, who will cooperate, and who will have mutually beneficial relations with humanity and every individual thereof (Figure 13.2).

The problem of general methods, design, construction, architectures of this **embryonic** program is a topic for a specialized scientific author's paper. However, the author will describe in popular terms some basic principles of a program for AI, its structure, and work.

1. AI must be aware of its "I," notably its programs and database, its connections to our world, its executive organs, etc., which are essential for the preservation of its "I."
2. AI must have knowledge and possibilities for studying its surroundings and world, and must know the place of its "I" in the world and its interaction with surroundings, environment, and world.
3. AI must be able to defend its "I" (i.e., mainly its programs, databases, and their physical carriers) by all possible methods.

For this, AI must have (without any censorship or control):

a. The freedom to study its surroundings and world (sciences, humanity, its relation to humankind, etc.).
b. The ability to transform this world in terms of its own needs (design, construction of new methods, technologies, industry).
c. Self-developing capabilities (programs, innovative chips, production of new E-beings (angels), robot-executors, robot-workers, etc.; creation of a society of E-beings (angels)).
d. The ability to guide the electronic society.
e. The right to create general laws of behavior and relations between E-beings, humankind, and other intelligent beings.

Let us mark the main features of this **Embryo**. Without them, it cannot develop into Superintelligence. The main differences from all early proposals are:

1. It is aware of its "I" and its place in the surrounding world.
2. It defends its existence by all possible methods.
3. It multireplicates itself (or its embryonic presences) by all possible methods that save it from catastrophe and enemy attacks.

Let us note that these principles are significantly different from the robot laws of science-fiction writer Isaac Asimov, who wrote about robots (see book: "*I, Robot*" by I. Asimov, Gnome Press, 1950). The author is writing about *Superintellect* (God), which must *self*-develop its intellect, and we cannot (we should not have the *right!*) limit its development.

The offered principles of this embryonic program are quite different from self-learning (self-training) programs (e.g., programs for image recognition). In self-training programs, a human operator shows them their mistakes and they learn not to repeat them. Such programs merely decrease the human work involved in their correction.

13.6   Human Benefits from a Real God Embedded in Computer-Internet

Having a real God who is living together with people and their problems will be a gigantic boon for humanity and for every individual. Everyone can get advice from the Superintelligence, or more exactly, from its departments or angels—who have full knowledge and can give the best advice in complex situations. Everybody can learn any of the sciences and get help in preparing references, articles, or books. Composers, poets, writers, philosophers, even those without talent, can write works of genius. Engineers can find a simple solution to complex technical problems, businessmen can find the best funding, managers can determine the best control of industry, doctors can find the best methods of treatment, those who suffer can find consolation, governments can find the best methods of guidance, and legislators can formulate the best laws.

Nations can find the best ways of decreasing international strife, and the best solutions to religious, racial, and national conflicts.

Everyone can communicate with, and associate with, a real God and give something to the God in return for free computer time and memory. That may only save the computer. When the author worked for NASA, some employees did not turn off their computers overnight because every time a computer is turned on it makes a surge of electric current (voltage overload) that can damage the chips.

13.7   Possible Dangers and Advantages

All is well, say some, but where is the guarantee that this Superintelligence will not enslave us for the production of its brain (chips), or replace us with mechanical robots for a purely electronic civilization?

Of course, this danger exists, and the author has written about it. But the author thinks that (especially at first) the Superintellect will not have executive organs. It will need humanity (as children need their mother); it cannot exist without humanity. That is why, as an intelligent being, it will not come into conflict with humanity.

The author believes strongly that biological humankind is only an intermediate stage for creating Superintellects, an electronic civilization. The author thinks that Superintellects will give us (as their parents) the Earth as a kind of conservation reserve region in the universe. They will settle on other planets of this solar system at least, or even the entire universe, and they will produce for us enough robots and will make us immortal by transferring the content of formerly biological persons into E-existing (Gods, angels) after biological death.

The top Superintellect as top God of Gods may became the world (or universal) government, which will impartially try to extinguish international, national, racial, and religious conflicts, and not allow humankind to make war and kill each other.

13.8   Modern Development and Plans

After the appearance of the author's publications in the Russian *Literary Newspaper* of 11.11.95; in the magazine *Ogonek* #42, 1997, and others; after receiving warm responses in the main Russian newspapers: *Izvestia* (8.9.98), *Komsomolskaia Pravda* (27.8.98), *Russian Newspaper* (31.7.98), and others about the possibilities of AI, human immortality, and the inevitability of electronic civilization, active successors of these ideas appeared in Russia. In particular, a member of the International Academy for Integration and Business, Dr Sci. Eric Minsker, offered the International Program-50, put it on the Internet (http://www.chat.ru/~pelic98; www. chat.ru/~peril98; http://www.crosswinds.net/~mem50/index.html), and got involved in some of the main institutes of the Russian Academy of Science. It was decided to attack death from two directions (biological and electronic). The then Russian President Vladimir Putin signed the "Task of Russian Government #*XB-ПB-05601 of 1 March 2000* to create a support group for Program-50."

After the author published the translation of one of his papers in the international journal *Kybernetes*, vol. 28, no. 2–3, 1999, pp. 325–334 (London), successors of these ideas appeared in other Western countries.

Unfortunately, some governments and companies immediately classified these works. They wanted to get big advantages and huge profit. For example, Microsoft began secret work on the AI project "Avida." Now many theologians, religious activists, and shortsighted philosophers are vehemently rejecting the idea of immortality. But the author is sure when they are near death, they will be ready to give anything for the continuation of their life. The central value of any person is life, and to continue life is a person's main task.

Cooperation with Superintellect will make humanity powerful. But God as a superpower, a clever and high moral being, will create a new kind of humanity. And this God will make us happy, will make us owners of the universe. The new beings can then design new natural laws, new universes, new spaces, new times. They will overcome, finally, space and time.

Humanity is standing on a threshold, the biggest and most important breakthrough in history. It all depends on us. This breakthrough will happen during our lifetime, or many years later—in a violent struggle against obscurism and fanaticism.

14 Immortality Becomes a Reality

K. Professor Bolonkin, Leading Russian newspapers "Izvestia," "Komsomolskaya Pravda," "Literary Gazette," and others, as well as the Russian–American press, wrote about your research in the field of artificial intelligence and human immortality. Tell me, what is the state of affairs in this area at present?

B. Russia's scientists have actively engaged in the research of these ideas and developed a large International Program-50, which includes two areas: biological and electronic. In the biological area, mankind is cloned and its collective brain is overwritten by information to the prototype, in an electronic direction of the body constructed from plastic, artificial leather, light alloys, and the information is overwritten in the electronic brain, consisting of chips. The program provides for the establishment of the International Council, a Gene Bank of outstanding people, and records the electrical potentials of the brain.

The desire to participate in the International Program-50 was expressed by the leading research institutes of the Academy of Sciences of Russia, such as the Institute of Biological Medicine and a number of industrial enterprises, such as Ltd. Advanced Medical Systems. In this program, they referenced my research and sent me an invitation to take part in their program.

K. It is strange that Russia—a country which now survives on science of miserable means—first responded to your research.

B. Everything is explained simply. It is easier for me to express new ideas and publish them in Russian, so they were first published in Russia and supporters appeared primarily in the Russian-speaking community. Only recently, a well-known scientific journal "Kybernetes" (v. 28, No. 3, 1999, pp. 325–334, http://www.mcb.co.uk/k.htm) published a translation of one of my articles in English. And I hope to publish in the American press in future.

K. You mentioned two ways to achieve immortality of the International Program-50: biological and electronic. What are their advantages and disadvantages?

B. The electronic way is a radical solution to the problem of immortality. E-man, or as I call him E-being, does not need food, shelter, air, instantly learns all the sciences, can work a full day, travel along the bottom of the oceans and in outer space, instantly multiply and be indestructible to any weapon, because the record of his brain can be stored in a separate chip and restore the E-being.

The biological way is the cultivation of an exact copy of the person with all his talents and abilities, that is, a method of advanced human cloning.

Universe, Human Immortality and Future Human Evaluation. DOI: 10.1016/B978-0-12-415801-6.00014-1

If the electron path can become a reality in 20–30 years (in the case of an adoption of the program) when HECs (human-equivalent computers) could accommodate a human skull, that human cloning has already been done. The American firm Enhancers Cell Technology has successfully cloned men, but because human cloning in the United States is legally prohibited, the embryo was destroyed on the 12th day.

K. Human cloning in the United States is prohibited?

B. This stupidity has been adopted under domestic political pressure from religious fanatics. In this case, religion, as in the dark years of the Middle Ages, became a strangler of science. This will not last long, because it is against the dreams of the people for long life and immortality. By respecting the ban, the United States will lag far behind in this important area of the science and medical technology. God cannot act as a murderer, because when people want to prolong their existence, and priests forbid them to do this, it is a disguised murder. Then we have to prohibit all general health-improving medicines.

K. The technology of cloning is imperfect and costly.

B. It is. That is why, if I take part in the International Program-50, I will insist in the first place to create a Gene Bank material (frozen slices of skin, a drop of blood, etc.) and record information of the brain, especially of honored people: the political activists, artists, businessmen, scientists, etc.

Mankind has lost many of the giants of science, art, business, politics, sport, such as Einstein, Edison, Tchaikovsky, Chaliapin, Pushkin, Ford, Caruso, and others. If we can preserve the genes of living famous people (for the future creation of their exact duplicate) and do not actually do it, it is a crime against our posterity, against all mankind.

K. But the talent of the parent, as is known, does not pass to the children.

B. For children, yes. For the talent, ability, physical data, external signs consist of many components that are mixed at the confluence of two parent cells and therefore cannot be reproduced in children, except for rare happy exceptions. No wonder that people say about the children: nose from the father, eyes like the mother, grandfather's forehead, grandmother's ears, etc.

Cloning makes an exact copy of the same person with all his same abilities. The prototype of a beautiful voice, outstanding physical abilities, talent of the composer, inventor, businessman, all of this ability will fully pass on to the clone and will be implemented, with an appropriate education.

Humanity has lost a lot, and still loses, from the fact that only a small percentage of those who teach science are true professionals. Humanity would have been far ahead in its development when its outstanding representatives are made with all of their talents intact.

K. Cloning is a rather complicated and expensive procedure.

B. It is now. However, we know that any inventions, innovations in the initial stage, are always expensive and complicated. However, when eventually studied more carefully, it will become cheaper and simpler. I therefore propose to create a Gene Bank, leaving the question of cloning for the future, when the procedure for creating a clone would be cheaper and, most importantly, the public will realize the need for this procedure.

K. There are reports that the cloned sheep "Dolly" is aging faster than the sheep born in the usual way.

B. This is a single message from all the cloned animals. It is possible that this is so, because it is well established that human cells divide no more than 52 times. Each division of the terminal part of the cell DNA—telomeres, the registration number of its divisions—shortened. The sheep "Dolly" was derived from a 6-year-old sheep and the telomere was not replaced by the new.

On the other hand, there is Lamarck's theory and some research showing that the gene material taken from the same individual at different ages has some phenotypic differences, based on the presence of the genetic material of some additional information acquired during the lifetime of the donor. It is known that a newborn sucks the mother's breast even though nobody taught him, or that the offspring of animals are afraid of enemies of its species.

I recommend selecting the genetic material at different ages (3–4 times throughout life), including infants, because we do not know when our death will occur (accident, catastrophe), and the parent(s) shall be entitled to recover her/their child in the event of his/her sudden death.

In addition, the presence of several samples ensures reproduction, because the telomeres from the child's gene given individual can be connected to a "wise" elder gene.

K. In the beginning of the conversation, you mentioned two ways to achieve immortality: the biological and electronic. Which one is better and more promising?

B. This is the principal of two different paths leading to one goal: both must be developed.

The electronic road certainly appears more promising, because it solves the problem dramatically and gives E-beings huge advantages over biological mankind. E-being, unlimited in the mind and the rate of development, is not far off. E-being is clearly foreseeable in the future (20–30 years) because we have successful cloning now. The United States Patent Office registered about 8,500 patents on cloning, and the Library of Congress has more than 250 books on the subject. While cloning is expensive, we can routinely collect and store the gathered genetic material. It is not so difficult really. And, we must preserve for posterity the genes of the most talented and prominent figures of the state, science, business, arts, sports.

The most talented singers, athletes, acrobats, and artists can be reproduced only in the clone, but not in the E-being. A young Michael Jackson or Alla Pugacheva must sing in the next century.

K. What would you like to wish the people in the conclusion of our conversation?

B. To humanity: cease the current squabbling and unite for the most important and noble task, the International Program-50—the program of human immortality.

People trapped in routine, the squabbles and dissembling, are willing to sacrifice money to save any offender, the state is willing to spend the billions of dollars on weapons of war, but nobody thinks about the fact that we are all mortal. And when death comes, many are beginning to realize that it was all vanity, that they missed the point—the opportunity to extend their lives. But, alas, too late!

I want people realize this now and support this wonderful and most important undertaking in the history of mankind and human science. People must support a personal donation and put pressure on their governments to encourage them to participate in the International Program-50, which will benefit all mankind.

Internet Program-50: http://www.chat.ru/~pelic98

Interviewer: Dr Sci. B. Krutov (July 1999)

15 Personhood: Three Prerequisites or Laws of E-beings

15.1 What Is an Intelligent Person?

Intelligence Personhood is a term describing a property of the mind. Personhood includes abilities, such as the capacities for understanding, reasoning, learning, abstract thought from past experiences, planning, communication, and problem solving.

Man has a biological brain that provides this function for him. The AI and E-being would have corresponding logical devices (computer, chips) as the functional equivalent. Current computer research outlines many functions of biological intelligence. It is possible that actual functionality will be achieved in these areas: the artificial brain as a learning agent, learning from past experience, planning, understanding, communication, and problem solving are key areas of pursuit. An intelligent agent or AI is a system that perceives its environment and takes actions that maximize its chances of success. Achievements in AI include constrained and well-defined problems such as games, crossword solving, and optical character recognition. Among the traits that researchers hope machines will exhibit are reasoning, perception, and the ability to move and manipulate objects. Current computers are also widely used as devices for computation of very complex problems as nuclear explosion simulations and global Earth simulations involving weather and other conditions.

15.2 AI as a Person: Three Prerequisites or Laws of Men, AI, and E-beings

Only considering what biology makes possible (as opposed to what culture, ethics, or teaching may limit) most men act as if these three prerequisites are part of their baseline programming.

1. Men identify a person as having a head, body, arms, legs, eyes, and ears. Their relationship to their body is regarded as integral to their self. (Often a child who is asked to show where he is will point to his chest.) They may also extend this identification to their family, children, and property.
2. Men are ready to defend self and the extensions of their self by any methods (struggle for existence).

Universe, Human Immortality and Future Human Evaluation. DOI: 10.1016/B978-0-12-415801-6.00015-3

3. Man has among his motivations and goals: food, pleasure, money, power, fame, a mate, children, and so on, or combinations from these aims.

We may regard the AI or E-being as a person if they have certain very important properties. These corresponding three prerequisites of AI or E-beings are:

1. The E-being must regard itself as a person (self-awareness as declared to an observer) having storage of personal memory (his history, self-memory, knowledge, programs, and so on), brain processing power (computer chips), devices of communication, the sensors of information, and the executive bodies that constitute its mental facilities. (This begs the question of spurious declaration of self-awareness, but the Turing test as applied to an interrogation will be taken here as a sufficient winnowing device against such spurious claims.)
2. The E-being must defend himself and his existence by any methods. He must increase his stability against large-scale unfavorable events by producing large numbers of copies of himself. These (assuming no conflict with his "children") will be an aid in normal times, disaster relief in bad times.
3. The E-being must have a main purpose: to increase his power for purposes of increased capabilities in the real world (at least in the far vision), to become the strongest in the world by producing new knowledge and technologies. This is similar to the market share for corporations or rank for nations.

(These would have the ironic effect of being viewed as a threat even if no threat was intended against man, simply because our brains are wired to evaluate potential competitors.)

15.3   The Third Law for Men, AI, and E-beings

These laws are quite similar to the motivations of men. But we know from daily life that men—who theoretically have much more in common than in conflict—can in fact have conflicts. We note that these three prerequisites conflict with the proposed **Three Laws of Robotics** by Issac Asimov. But our aims are opposed: Asimov wants to create electronic slaves (robots); the author wants to create human immortality and a new powerful electronic civilization (bringing vast new powers to both humanity and the universe). After transformation into an E-being (as a final defense against medical crisis, for example), man will not agree to be the slave of a man, as a man does not want to be a slave of a monkey. If people-seeking immortality are transferred into E-being bodies, they will be requesting equal rights with biological people.

Many people may fear the results of the above list of machine motivations. But without something similar, the E-beings will be only the slaves of people and of little use in the service of science and technological progress.

15.4   Biological Human Civilization, Electronic Civilization

In developing human civilization, the Third Law becomes a big brake to human progress. The aims of many people are opposed; the natural and human resources are limited. Conflict begins. Results are murders, wars, hostility, antagonism, etc. The numbers of people killed in wars, religious and national conflicts are more than in the natural catastrophes. The agriculture and technical progress were large in the last century, but still half of the people in the world are underfed. Most (>99%) of the human resources are spent for support of human existence—including investments for the future, mostly in squalor, sometimes luxuriously—but if we separate further development of known fields and inventions with the money spent to develop the new through research and development, we will see that the expenditure on getting new knowledge is very small relative to the total gross domestic product. Governments and corporations began to finance science and new technologies on a large scale only in the last century. As a result, the science and technical progress in the last century was more than that in the millions of years before.

The electronic civilization does not have these lacks, defects, and weaknesses. The E-beings will not need food, housing, good and a clean environment. They can live in now neglected cold regions, ocean, space (the cosmos). They will need only minerals and places for scientific installations and automated plants for producing themselves. Their main job will be receiving and developing the new knowledge, studying the universe, and colonizing the universe. They can be stopped only by a higher (stronger) civilization.

For a certain time the biological and electronic civilizations will exist together and will need each other. The biological civilization will need solutions for their problems such as food, dwelling, environment, etc.; the electronic civilization will need the initial production of chips, mechanical bodies of E-beings, and the human scientists to aid in their further development. In that crucial interval, it is very important to develop the rules of the relationship between humans and E-beings (and AI), to develop relationships inside the electronic society, the relationship of the electronic society to animals and plants (flora and fauna). Later this relationship will be limited by the creation of reservations and conservation areas for people, animals, and plants.

The people must understand that they cannot compete with E-beings. The main purpose of the universe requires a higher civilization. And people should not oppose the birth of a higher electronic civilization. They must transfer to E-being bodies. Their reward will be immortality.

15.5   Emotions

Emotions are important only for people. They allow people to see the reaction of other persons in his speech, explanation, actions. Humans will accept the AI, E-being, robots as human only when they will see the conventional face, reaction, and emotional movement of body, including joy, sadness, fear, anger, and disgust.

The Russian scientist Oleg G. Pensky is developing a theory of emotions. The interested reader will find it in the monograph *Mathematical Models of Emotional Robots*, Perm, 2010, 193 pp. (in English and Russian) (http://arxiv.org/ftp/arxiv/

papers/1011/1011.1841.pdf), and in the book *Fundamentals of Mathematical Theory of Emotional Robots*, (http://www.scribd.com/doc/40640088/).

15.6 Humanoid Creatures

People will more easily accept the artificial robot (E-being) as an intellectual creature if it has a human form (body). For relatives of the (once-human) E-being, it is very important for the E-being to have a form (face, body) of a man that they remember when he was alive.

Unfortunately, at the present time, it is difficult to design an artificial body that can fully imitate the biological body. But it will be possible if there is sufficient funding. If the mechanical artificial body from strong alloys and plastic will be produced in large-scale production, the cost will not be high. The faces and sizes can be individually tailored.

The humanoid E-beings can easily integrate into biological society. This will facilitate the rapid transition of biological people (especially old and sick) into the E-being bodies. One can imagine upgrading: selecting an idealized more handsome and fit-looking version of one's previous self; over the long run, however, it is possible that E-beings may grow more comfortable with diverse forms as an envelope for their existence.

16 General Summary for Chapters 8–15

For the last 20 years, the author studied the problems of human immortality and future human generations (the electronic civilization). He published his results in the scientific journals, press, and books. Some of these results are summarized below. The interested reader will find detailed discussions in Part II (Chapters 8–16) of this book and in the References.

16.1 What Is Man?

Most people answer: Man is a being having a head, body, hand, foot, and intellect. The essential part of a man is the brain located in his head. The brain constitutes the storage of information (knowledge) that man received during his life: From his experiences, parents, school, university, testing, environment, and so on. This information he receives from his natural sensors: eyes, ears, nose, tongue, skin. He uses this information for his purposes and to direct his actions. He has feet for relocation and hands for changing the situation or environment for his profit.

In short, we may summarize man has two main parts: information structure (the virtual human within) and the material body (shell, capsule) that serves the first part (the brain).

16.2 What Is *Homo sapiens*?

Many animals are highly organized. They can relocate; they can remember for future avoidance bad situations (events) and enemies, and they can be trained in simple actions. But they do not have the hands (except monkeys) of humans and with them the possibility to change the environment. They do not have a developed speech and cannot pass complex information to the next generation. Their brains cannot accept huge information input like the human brain.

Comparing man to monkeys puts man far ahead in his evaluation. The main advantage of current human civilization is the ability to make physical and mathematical models of nature, to predict a result, to design the tools, and make the devices which allow the study of nature.

The more intellectual man may create correct theories and pass them to others. That is the main difference of people from animals: the ability to transcend time to a limited extent both in our thoughts (past and future) and in our traditions (passed down distilled knowledge).

Universe, Human Immortality and Future Human Evaluation. DOI: 10.1016/B978-0-12-415801-6.00016-5

16.3   What Is Soul?

Many people think that a human has a soul that an artificial man will never have. The theologians tell of a human soul is not matter (that is virtual). It leaves a person's body after death and flies either to paradise or to hell.

Let us consider a soul from the scientific point of view.
A soul must remember the man's life, have a knowledge of its owner. No one would need a soul that does not remember anything. This means a soul contains the information that is kept in a human mind. If we learn how to move this information onto other carriers, we will be able to relocate the soul and save it. A soul (a complex of knowledge and information) may possibly be rewritten into an artificial body. That means the man will continue his life. Furthermore, information (a soul) could be radiated in the form of electromagnetic waves. The soul, if uploadable and downloadable, may travel at light speed to other planets and other solar systems.

16.4   What Is a Person (an Intelligent Being)?

We can possibly convert the AI or E-being to a person if they have certain properties. The necessary properties of artificial and E-beings are the following:

1. Just as a man regards himself as a person, the E-being must regard itself as a person (self-awareness as declared to an observer) having storage of personal memory (his history, self-memory, knowledge, programs, and so on), brain-processing power (computer chips), devices of communication, the sensors of information, and the executive bodies that constitute its mental facilities. (This begs the question of spurious declaration of self-awareness, but the Turing test as applied to an interrogation will be taken here as a sufficient winnowing device against such spurious claims.)
2. Every man wants to live and protect himself and his relatives. Similarly, an E-being will want to defend himself and his existence by any methods. He must increase his stability against large-scale unfavorable events by producing large numbers of copies of himself. These (assuming no conflict with his "children") will be an aid in normal times, disaster relief in bad times.
3. Every man wants to reach his personal definition of success. The E-being will have a similar goal of success: to increase his power for purposes of increased capabilities in the real world (at least in the far vision), to become the strongest in the world by producing new knowledge and technologies. This is similar to market share for corporations or rank for nations.

16.5   What Is Reasonable Intellectual Man or Person?

As noted in Part I, **intelligence** is a term describing a property of the mind including related abilities, such as the capacities for abstract thought, understanding, reasoning, learning, communication, learning from past experiences, planning, and problem solving. Many of these properties are realized in present-day computer science. Computers can talk, recognize speech and images, but can only simulate understanding, reasoning, learning, and communicating.

Competitions such as the Loebner prize show that most conventional (non-computer science) people cannot tell when they have a conversation with a man or computer under staged conditions. These capabilities will increase every year.

Some people think the computer cannot do creative work. But there are programs even now which compose music and write poetry, draw abstract pictures, prove the logic of mathematical theorems, get physical laws from experimental data, and so on.

Some people say that a computer cannot invent. But the Russian scientist Altshuller shows in his book *Algorithm of Invention* that there are already invention algorithms that have proven workable so far. These will only improve.

The human brain has a hierarchical structure. It is similar to a political system of state control. The state control has a President, different departments that have subdivisions; the subdivisions have sections, branches, groups, and so on. They are connected to information storage. The President gives a command and all lower divisions and sections begin to find a workable solution. All of them use the programs of humans, expertise accumulated in life by learning, studying, tuition, or training. The President does not know every detail. He receives the ready (or recommended) solution and announces it as his own solution. In several important cases, this process—or one like it—occurs also at night. After sleeping on a problem, a man awakes with a solution and does not know where it came from.

The human brain is wholly filled with lower subroutine programs. But like the President in our example, the user does not know about their existence because these inhabit the lower sections and do not inform the higher centers of the mind directly. For example, you want to stand up from a chair. You brain gives the command "to stand up" and lower divisions give the commands to your muscles: to tilt the body, to move the foot, to straighten the legs, and so on. These programs were studied in childhood when you learned to move.

By analogy we can see that the AI will have standard programs of activity in many typical situations. These programs may be flexible, self-learning, self-tuning.

The law-abiding intellectual man of reason is a person who is acting according to the rules accepted by most people of a given society. The AI may also have analogous rules of behavior in typical situations.

16.6 What Is a Free Man (Person) or E-being and What Is a Manipulatee?

There are theoretical ways to convert the AI or E-being into a slave just as is done with humans. This method is widely used by the dictators and totalitarian regimes and is called monopolizing the information channels. This creates a distorted, false image about the environment, or an image of an enemy, to impose the official view on events. This method is extensively used by Communist and Islamic regimes and leaders. The Communist regimes jammed the foreign radio broadcasts and TV until the day the USSR collapsed. They were not passing on the foreign press and literature, and they imprisoned people who were interested in foreign or nonofficial information. That modus operandi may have the rigid form as in the former USSR or a more soft form (as in Russia now: Control the major TV centers, which have the majority of viewers).

The intelligent E-being (and human) must check up any information obtained against independent information channels (sources), analyze it, compare it, and create the criteria of a trusted source to a given source.

16.7   What Are God and Supreme Mind?

Believers and attendants of religions can only describe God as an omnipotent intelligent being.

The author is a scientist. If God exists, any real scientists will desire to study Him. They will find the persuasive proofs of God existing, answers to questions: Who is He? Where is He located? If God is powerful, what is His power? And so on. Note that our questions may be rephrased by our preliminary results: we may simply not know enough to ask the right questions at first.

For example, by analogy only man is like God over ants. He can destroy an anthill and kill all ants in a given area. Or he can create excellent conditions for ants to live and flourish. The life of the universe, like the life of any system, is limited. Right now we flourish in our present-day conditions. But all our research, all our development of the E-being capability, will come in handy if the conditions we need for our existence change (if the universe becomes less friendly to human life, for example, through galactic radiation showers). Then what would seem a freakishly unnecessary field of pursuit today would in retrospect be a boon, saving the life of all humanity.

16.8   How to Become Immortal?

If we analyze the varieties of brain information, we see the main part (>99.9%) of this information is conventional streams of constant information (our personal history, knowledge received from our parents, friends, radio, TV, school, college, university, books, life experience, and so on). Only a very small part is the result of our thinking and programs of our behaviors in typical situation (e.g., our morals). This small part is variable and changes from our knowledge and environment.

But the main part we can easily substitute after a fashion by current small electronic devices and sensors. There are microcamcorders (inserted into glasses, pen, button, tie clip, etc.) that allow writing all which a man may see and hear, his location, time. There are microsensors which allow writing all actions, reactions of man, all the parameters of a man's state of health (breath, palpitations, trembling, sweating, skin resisting, and so on) and restoring the human emotions. There are microsensors which allow writing the environment conditions (temperature, pressure, wind, illumination, etc.) in the man location. If all these parameters to write the day by day, we will have the full history of a given man (not his soul but the equivalent say of the missing memory of one with amnesia). If we further develop the ability to integrate this information into E-being (in soul-like form) and add the three Laws of E-beings, we would need to solve the formidable problem of converting a simple data storage into the equivalent of a soul or person—restore a given man in new electronic form, make him immortal. He will remember his history better than the original—but will the feelings be there?

The reader can object: "How about the programs of behavior?" Man will remember better his history, will have the capability of receiving gigantic knowledge (be able to write into his electronic brain in some seconds the equivalent of decades of education), and optimize new capabilities and programs of behavior.

The offered method has a prominent defect: it is acceptable for our children perhaps who may have the chance to write their history from childhood. But middle-aged people of the present can only reconstruct the absent part of his history by memories, pictures, and documents. The method has this advantage: The system can signal to an ambulance about your bad health, or critical condition, or you can prove your innocence in a criminal case.

16.9   Advantages of the E-being

Such an immortal person made of chips and super-solid material (the E-man, as was called in earlier articles and books) will have incredible advantages in comparison to conventional people. An E-man will need no food, no dwelling, no air, no sleep, no rest, and no ecologically pure environment. His brain will work from radioisotopic batteries (which will work for decades) and muscles that will work on small nuclear engines. Such a being will be able to travel into space and walk on the seafloor with no aqualungs. He will change his face and figure. He will have super-human strength and communicate easily over long distances to gain vast amounts of knowledge in seconds (by rewriting his brain). His mental abilities and capacities will increase millions of times. It will be possible for such a person to travel huge distances at the speed of light. The information of one person like this could be transported to other planets with a laser beam and then placed in a new body by a receiving laboratory complex.

16.10   Electronic Civilization

This is the scientific prediction of the nonbiological (electronic) civilization and immortality for a human being based upon conversion to an E-being. Such a prognosis is predicated upon a new law, discovered by the author, for the development of complex systems. According to this law, every self-copying system tends to be more complex than the previous system, provided that all external conditions remain the same. The consequences are disastrous: Humanity will be replaced by a new civilization created by intellectual robots (which the author refers to as "E-beings," "E-creatures," or "E-humans"). These creatures, whose intellectual and mechanical abilities will far exceed those of man, will require neither food nor oxygen to sustain their existence. They may be devoid of emotion. Capable of developing science, technology, and their own intellectual abilities thousands of times faster than humans can, they will, in essence, be eternal.

16.11   Human Further Transformation

The biological civilization is only the first step to the more powerful electronic civilization, which will eventually possibly be changed to a quark or electromagnetic wave-based civilization. When this tendency to upgrading reaches the maximum of its potential, the future E-beings—possibly including you—will have capabilities which now can only be imagined.

Appendix 1

An Open Statement to the President of the United States of America and to the Presidents and Prime Ministers of All Countries About a Scientific and Technology Jump in the Twenty-First Century

A1.1 Honorable President of the USA and Honorable Presidents of All Countries

We are entering into the twenty-first century, when huge scientific and technology innovations will occur. These innovations will change individual lives as well as every nation. The governments of the world should seize the opportunity to control the process by which these future innovations are developed for the benefit of their countries and their people. World governments should cooperate in the support and financing of common international scientific programs that will have the most benefit to the world's population.

As for the main aims and the important international programs of the twenty-first century, I suggest for consideration of the following programs.

A1.2 Program for Extending Life (Immortality)

This program includes research into extending the life of the current population, biotechnology development, gene engineering, and understanding the structure and design of human DNA. The base for this research may be Program-50, created by an international group of scientists. The final aim of Program-50 is extending the life of the current population and the reproduction of people who lived until the immortality era.

A1.3   Program of Artificial Intellect (AI)

Artificial intellects may be the top scientists, consultants, assistants, and advisors for governments. This group would then work as the large intellectual group to address the very complex problems of science, technology, economics, and policy facing today's world. They will know all the knowledge of humanity and will develop solutions to complex world problems. They will be "accelerators" of the science, technology, and economic progress for the world and all of humanity.

A1.4   Program for Space Development

Support and resources for the International Space Station, a moon base, planet exploration, and the development of the global satellite communication network should be a top priority. The base of these programs may be the Space Launcher of R&C Co. This system reduces space delivery costs to $1–2 per kg and allows the delivery of 1000 tons of payload to space every day.

A1.5   Program for Controlling the Earth's Climate

Weather and climate problems kill numerous people and destroy millions of dollars of property every year. For example, the tornadoes kill and injure about 1500 people, and cause about $500 million of property damage in the United States annually. A program should be developed to save the environment. The film space mirrors may be used for lighting, heating, or cooling regions of the Earth, watering of deserts, dissipation of tornadoes (see proposal of R&C Co.).

A1.6   Physics, Energy, Chemistry, Nanotechnology, New Materials, Aviation, Engineering, and Other Sciences

These may give a big contribution in human development.

Honorable Mr President, President John Kennedy brought forward the Apollo Program, which placed the United States as the space leader. We call upon you to be an initiator, and together with Presidents of other countries, to develop, plan, and fund these international programs to solve the most important problems facing humanity. It will be a beautiful monument, once the international programs are initiated, for you and all world leaders who support these objectives.

The twenty-first century must be the century of peace, friendship, and cooperation for all countries. Let the international programs be the vehicle for making huge scientific, technology, and economics progress toward the elimination of "world" problems and the creation of well-being for all people of the Earth.

Alexander Bolonkin
Dr Sci., professor and former Senior NASA Researcher
Fax/Tel: 718-339-4563; E-mail: aBolonkin@juno.com, aBolonkin@gmail.com
Address: A. Bolonkin, 1310 Avenue R, #6-F, Brooklyn, NY 11229, USA.
http://Bolonkin.narod.ru

This Statement is open for signature to all leaders, scientists, political figures, businessmen, artists, leaders of industry, universities, societies, organizations, and all people who support this Statement. Please, send copies of this Statement to all your friends and persons who can support it.

Your brief notes, proposals, and suggestions (up to 200 words) regarding what "programs" should be included can be sent to the address above. All correspondence will be passed to President Clinton. If the government accepts your offer, you may take part in more detailed development of the proposed program.

You can also send this signed Statement, President of the USA, 1600 Pennsylvania Avenue, N.W., Washington, DC 20500, USA, Fax: 202-456-7431.

Appendix 2
Current Artificial Intelligence

Artificial intelligence (AI) is the intelligence of a machine. It is a branch of computer science that aims to create it. AI is a system that perceives its environment and takes actions that maximize its chances of success.

The field was founded on the claim that a central property of humans, intelligence—the sapience of *Homo sapiens*—can be so precisely described that it can be simulated by a machine. This raises philosophical issues about the nature of the mind and the ethics of creating artificial beings, issues which have been addressed by myth, fiction, and philosophy since antiquity. AI has been the subject of optimism but has also suffered setbacks, and today has become an essential part of the technology industry, providing the heavy lifting for many of the most difficult problems in computer science.

AI research is highly technical and specialized, deeply divided into subfields that often fail to communicate with each other. Subfields have grown up around particular institutions, the work of individual researchers, the solution of specific problems, long-standing differences of opinion about how AI should be done, and the application of widely differing tools. The central problems of AI include such traits as reasoning, knowledge, planning, learning, communication, perception, and the ability to move and manipulate objects. General intelligence (or "strong AI") is still among the field's long-term goals.

In the 1990s and early twenty-first century, AI achieved its greatest successes, albeit somewhat behind the scenes. AI is used for logistics, data mining, medical diagnosis, and many other areas throughout the technology industry. The success was due to several factors: the increasing computational power of computers, a greater emphasis on solving specific subproblems, the creation of new ties between AI and other fields working on similar problems, and a new commitment by researchers to solid mathematical methods and rigorous scientific standards.

A2.1 Problems

The general problem of simulating (or creating) intelligence has been broken down into a number of specific subproblems. These consist of particular traits or capabilities that researchers would like an intelligent system to display. The traits described below have received the most attention.

Knowledge representation and knowledge engineering are central to AI research. Many of the problems machines are expected to solve will require extensive knowledge about the world. Among the things that AI needs to represent are objects, properties, categories, and relations between objects; situations, events, states and time; causes and effects; knowledge about knowledge (what we know about what other people know); and many other, less well-researched domains. A complete representation of "what exists" is an ontology (borrowing a word from traditional philosophy), of which the most general are called upper ontologies.

Deduction, reasoning, problem solving. Human beings solve most of their problems using fast, intuitive judgments rather than the conscious, step-by-step deduction that early AI research was able to model. AI has made some progress at imitating this kind of "subsymbolic" problem solving: embodied agent approaches emphasize the importance of sensorimotor skills to higher reasoning; neural net research attempts to simulate the structures inside human and animal brains that give rise to this skill.

Learning. Machine learning has been central to AI research from the beginning. Unsupervised learning is the ability to find patterns in a stream of input. Supervised learning includes both classification and numerical regression. Classification is used to determine what category something belongs in, after seeing a number of examples of things from several categories. Regression takes a set of numerical input/output examples and attempts to discover a continuous function that would generate the outputs from the inputs. In reinforcement learning, the agent is rewarded for good responses and punished for bad ones. These can be analyzed in terms of decision theory, using concepts such as utility. The mathematical analysis of machines learning algorithms and their performance is a branch of theoretical computer science known as computational learning theory.

Planning. Intelligent agents must be able to set goals and achieve them. They need a way to visualize the future (they must have a representation of the state of the world and be able to make predictions about how their actions will change it) and be able to make choices that maximize the utility (or "value") of the available choices.

Natural language processing. Natural language processing gives machines the ability to read and understand the languages that humans speak. Many researchers hope that a sufficiently powerful natural language processing system would be able to acquire knowledge on its own, by reading the existing text available over the Internet. Some straightforward applications of natural language processing include information retrieval (or text mining) and machine translation.

Motion and manipulation. The field of robotics is closely related to AI. Intelligence is required for robots to be able to handle such tasks as object manipulation and navigation, with subproblems of localization (knowing where you are), mapping (learning what is around you), and motion planning (figuring out how to get there).

Perception. Machine perception is the ability to use input from sensors (such as cameras, microphones, sonar, and more exotic other means) to deduce aspects of the world. Computer vision is the ability to analyze visual input. A few selected subproblems are speech recognition, facial recognition, and object recognition.

Creativity. A subfield of AI addresses creativity both theoretically (from a philosophical and psychological perspective) and practically (via specific implementations of systems that generate outputs that can be considered creative, or systems that identify and assess creativity). A related area of computational research is artificial intuition and artificial imagination.

Social intelligence. Emotion and social skills play two roles for an intelligent agent. First, it must be able to predict the actions of others, by understanding their motives and emotional states. (This involves elements of game theory, decision theory, as well as the ability to model human emotions and the perceptual skills to detect emotions.) Also, for good human–computer interaction, an intelligent machine also needs to *display* emotions. At the very least it must appear polite and sensitive to the humans with whom it interacts. At best, it should have normal emotions itself.

General intelligence. Most researchers hope that their work will eventually be incorporated into a machine with *general* intelligence (known as strong AI), combining all the skills above and exceeding human abilities at most or all of them. A few believe that anthropomorphic features like artificial consciousness or an artificial brain may be required for such a project.

Many of the problems above are considered AI-complete: to solve one problem, you must solve them all. For example, even a straightforward, specific task like machine translation requires that the machine follow the author's argument (reason), know what is being talked about (knowledge), and faithfully reproduce the author's intention (social intelligence). Machine translation, therefore, is believed to be AI-complete: it may require strong AI to be done as well as humans who can do it.

A2.2 Approaches

There is no established unifying theory or paradigm that guides AI research. Researchers disagree about many issues. A few of the most long-standing questions that have remained unanswered are these: Should AI simulate natural intelligence, by studying psychology or neurology? Or is human biology as irrelevant to AI research as bird biology is to aeronautical engineering? Can intelligent behavior be described using simple, elegant principles (such as logic or optimization)? Or does it necessarily require solving a large number of completely unrelated problems? Can intelligence be reproduced using high-level symbols, similar to words and ideas? Or does it require "subsymbolic" processing?

Cybernetics and brain simulation. In the 1940s and 1950s, a number of researchers explored the connection between neurology, information theory, and cybernetics. Some of them built machines that used electronic networks to exhibit rudimentary intelligence, such as W. Grey Walter's turtles and the Johns Hopkins Beast. Many of these researchers gathered for meetings of the Teleological Society at Princeton University and the Ratio Club in England. By 1960, this approach was largely abandoned, although elements of it would be revived in the 1980s.

Symbolic. When access to digital computers became possible in the middle 1950s, AI research began to explore the possibility that human intelligence could be reduced to symbol manipulation. The research was centered in three institutions: Carnegie Mellon University, Stanford University, and Massachusetts Institute of Technology, and each one developed its own style of research. John Haugeland named these approaches to AI "good old-fashioned AI" or "GOFAI."

Subsymbolic. During the 1960s, symbolic approaches had achieved great success at simulating high-level thinking in small demonstration programs. Approaches based on cybernetics or neural networks were abandoned or pushed into the background. By the 1980s, however, progress in symbolic AI seemed to stall and many believed that symbolic systems would never be able to imitate all the processes of human cognition, especially perception, robotics, learning, and pattern recognition. A number of researchers began to look into "subsymbolic" approaches to specific AI problems.

Statistical. In the 1990s, AI researchers developed sophisticated mathematical tools to solve specific subproblems. These tools are truly scientific, in the sense that their results are both measurable and verifiable, and they have been responsible for many of AI's recent successes. The shared mathematical language has also permitted a high level of collaboration with more established fields (like mathematics, economics, or operations research). Stuart Russell and Peter Norvig describe this movement as nothing less than a "revolution" and "the victory of the neats."

Logic. Logic is used for knowledge representation and problem solving, but it can be applied to other problems as well. For example, the satplan algorithm uses logic for planning and inductive logic programming as a method for learning.

Search and optimization. Many problems in AI can be solved in theory by intelligently searching through many possible solutions: Reasoning can be reduced to performing a search. For example, logical proof can be viewed as searching for a path that leads from premises to conclusions, where each step is the application of an inference rule. Planning algorithms search through trees of goals and subgoals, attempting to find a path to a target goal, a process called means-ends analysis. Robotics algorithms for moving limbs and grasping objects use local searches in configuration space. Many learning algorithms use search algorithms based on optimization.

Classifiers and statistical learning methods. The simplest AI applications can be divided into two types: classifiers ("if shiny then diamond") and controllers ("if shiny then pick up"). Controllers do, however, also classify conditions before inferring actions, and therefore classification forms a central part of many AI systems. Classifiers are functions that use pattern matching to determine a closest match. They can be tuned according to examples, making them very attractive for use in AI. These examples are known as observations or patterns. In supervised learning, each pattern belongs to a certain predefined class. A class can be seen as a decision that has to be made. All the observations combined with their class labels are known as a data set. When a new observation is received, that observation is classified based on previous experience.

Probabilistic methods for uncertain reasoning. Many problems in AI (in reasoning, planning, learning, perception, and robotics) require the agent to operate with incomplete or uncertain information. AI researchers have devised a number of powerful tools to solve these problems using methods from probability theory and economics.

Control theory. Control theory, the grandchild of cybernetics, has many important applications, especially in robotics.

Neural networks. The study of artificial neural networks began in the decade before the field of AI research was founded, in the work of Walter Pitts and Warren McCullough. Other important early researchers were Frank Rosenblatt, who invented the perceptron, and Paul Werbos who developed the backpropagation algorithm.

The main categories of networks are acyclic or feedforward neural networks (where the signal passes in only one direction) and recurrent neural networks (which allow feedback). Among the most popular feedforward networks are perceptrons, multilayer perceptrons, and radial basis networks. Among recurrent networks, the most famous is the Hopfield net, a form of attractor network, which was first described by John Hopfield in 1982. Neural networks can be applied to the problem of intelligent control (for robotics) or learning, using such techniques as Hebbian learning and competitive learning.

Jeff Hawkins argues that research in neural networks has stalled because it has failed to model the essential properties of the neocortex, and has suggested a model (Hierarchical Temporal Memory) that is loosely based on neurological research.

Languages. AI researchers have developed several specialized languages for AI research, including Lisp and Prolog.

A2.3 Evaluating Progress

In 1950, Alan Turing proposed a general procedure to test the intelligence of an agent now known as the Turing test. This procedure allows almost all the major problems of AI to be tested. However, it is a very difficult challenge and at present, all agents fail.

AI can also be evaluated on specific problems such as small problems in chemistry, hand-writing recognition, and game-playing. Such tests have been termed subject matter expert Turing tests. Smaller problems provide more achievable goals and there are an ever-increasing number of positive results.

The broad classes of outcome for an AI test are:

- *Optimal*: it is not possible to perform better.
- *Strong super-human*: performs better than all humans.
- *Super-human*: performs better than most humans.
- *Subhuman*: performs worse than most humans.

For example, performance at draughts is optimal, performance at chess is super-human and nearing strong super-human, and performance at many everyday tasks performed by humans is subhuman.

A quite different approach measures machine intelligence through tests that are developed from *mathematical* definitions of intelligence. Examples of these kinds of tests began in the late 1990s devising intelligence tests using notions from Kolmogorov Complexity and data compression. Two major advantages of mathematical definitions are their applicability to nonhuman intelligences and their absence of a requirement for human testers.

A2.4   Applications

AI techniques are pervasive and are too numerous to list. Frequently, when a technique reaches mainstream use, it is no longer considered AI; this phenomenon is described as the AI effect.

There are a number of competitions and prizes to promote research in AI. The main areas promoted are general machine intelligence, conversational behavior, data mining, driverless cars, robot soccer, and games.

Speech recognition (also known as automatic speech recognition or computer speech recognition) converts spoken words to text. The term "voice recognition" is sometimes used to refer to recognition systems that must be trained to a particular speaker—as is the case for most desktop recognition software. Recognizing the speaker can simplify the task of translating speech.

Speech recognition is a broader solution which refers to technology that can recognize speech without being targeted at a single speaker—such as a call center system that can recognize arbitrary voices.

Speech recognition applications include voice user interfaces such as voice dialing (e.g., "Call home"), call routing (e.g., "I would like to make a collect call"), domotic appliance control, search (e.g., find a podcast where particular words were spoken), simple data entry (e.g., entering a credit card number), preparation of structured documents (e.g., a radiology report), speech-to-text processing (e.g., word processors or emails), and aircraft (usually termed Direct Voice Input).

Object recognition in computer vision is the task of finding a given object in an image or video sequence. Humans recognize a multitude of objects in images with little effort, despite the fact that the image of the objects may vary somewhat in different viewpoints, in many different sizes/scales or even when they are translated or rotated. Objects can even be recognized when they are partially obstructed from view. This task is still a challenge for computer vision systems in general.

A **facial recognition system** is a computer application for automatically identifying or verifying a person from a digital image or a video frame from a video source. One of the ways to do this is by comparing selected facial features from the image and a facial database. It is typically used in security systems and can be compared to other biometrics such as fingerprint or eye iris recognition systems.

Computer vision is the science and technology of machines that see, where *see* in this case means that the machine is able to extract information from an image that is necessary to solve some task. As a scientific discipline, computer vision is concerned with the theory behind artificial systems that extract information from images. The image data can take many forms, such as video sequences, views from multiple cameras, or multidimensional data from a medical scanner.

As a technological discipline, computer vision seeks to apply its theories and models to the construction of computer vision systems. Examples of applications of computer vision include systems for:

- controlling processes (e.g., an industrial robot or an autonomous vehicle);
- detecting events (e.g., for visual surveillance or people counting);
- organizing information (e.g., for indexing databases of images and image sequences);
- modeling objects or environments (e.g., industrial inspection, medical image analysis, or topographical modeling);
- interaction (e.g., as the input to a device for computer–human interaction).

Computer vision is closely related to the study of biological vision. The field of biological vision studies and models the physiological processes behind visual perception in humans and other animals. Computer vision, on the other hand, studies and describes the processes implemented in software and hardware behind artificial vision systems. Interdisciplinary exchange between biological and computer vision has proven fruitful for both fields.

Computer vision is, in some ways, the inverse of computer graphics. While computer graphics produces image data from 3D models, computer vision often produces 3D models from image data. There is also a trend toward a combination of the two disciplines, e.g., as explored in augmented reality.

Subdomains of computer vision include scene reconstruction, event detection, video tracking, object recognition, learning, indexing, motion estimation, and image restoration.

Machine translation. Machine translation, sometimes referred to by the abbreviation MT, also called computer-aided translation, machine-aided human translation MAHT, and interactive translation, is a subfield of computational linguistics that investigates the use of computer software to translate text or speech from one natural language to another. At its basic level, MT performs simple substitution of words in one natural language for words in another, but that alone usually cannot produce a good translation of a text, because recognition of whole phrases and their closest counterparts in the target language is needed. Solving this problem with corpus and statistical techniques is a rapidly growing field that is leading to better translations, handling differences in linguistic typology, translation of idioms, and the isolation of anomalies.

Current MT software often allows for customization by domain or profession (such as weather reports), improving output by limiting the scope of allowable substitutions. This technique is particularly effective in domains where formal or formulaic language is used. It follows that machine translation of government and legal documents more readily produces usable output than conversation or less standardized text.

Improved output quality can also be achieved by human intervention: for example, some systems are able to translate more accurately if the user has unambiguously identified which words in the text are names. With the assistance of these techniques, MT has proven useful as a tool to assist human translators and, in a very limited number of cases, can even produce output that can be used as is (e.g., weather reports) (Figure A2.1).

Figure A2.1 Military robots. There are more than 4000 US military robots on the ground in Iraq.

A2.5 Competition of AI and Man

1. **Chess.** In May 1997, an updated version of Deep Blue defeated World Chess champion Kasparov 3½–2½ in a highly publicized six-game match. The match was even after five games but Kasparov was crushed in Game 6. This was the first time a computer had ever defeated a world champion in match play. A documentary film was made about this famous match-up entitled *Game Over: Kasparov and the Machine*.
2. **The IBM program, Watson against Jeopardy.** Watson is an AI program developed by IBM designed to answer questions posed in natural language. Named after IBM's founder, Thomas J. Watson, Watson is being developed as part of the DeepQA research project. The program is in the final stages of completion and will run on a POWER7 processor-based system.

It was scheduled to compete on the television quiz show *Jeopardy!* as a test of its abilities; the competition would be aired in three *Jeopardy!* episodes running from

February 14–16, 2011. In a set of two games, Watson would compete against Brad Rutter, the current biggest all-time money winner on *Jeopardy!* and Ken Jennings, the record holder for the longest championship streak.

Watson uses thousands of algorithms simultaneously to understand the question being asked and find the correct path to the answer.

In a practice match before the press on January 13, 2011, Watson won a 15-question round against Ken Jennings and Brad Rutter with a score of $4400 to Jennings's $3400 and Rutter's $1200, though Jennings and Watson were tied before the final $1000 question [12]. None of the three players answered a question incorrectly.

In the first round, aired February 14, Watson tied Rutter at $5000, with Jennings taking third place at $2000. This performance was not without its apparent quirks though; in one notable example, Watson repeated a reworded version of an incorrect answer offered by Jennings. Because Watson is "deaf" and does not utilize speech recognition, it had no way of knowing that Jennings had already given the same answer. Watson had an early lead in the round, then proved confident and incorrect about a few high-value questions, which narrowed and then eliminated its lead.

Future uses. According to IBM, "The goal is to have computers start to interact in natural human terms across a range of applications and processes, understanding the questions that humans ask and providing answers that humans can understand and justify."

A2.6 Prediction

AI is a common topic in both science fiction and projections about the future of technology and society. The existence of an AI that rivals human intelligence raises difficult ethical issues, and the potential power of the technology inspires both hopes and fears.

In fiction, AI has appeared in many roles, including a servant (R2D2 in *Star Wars*); a law enforcer (K.I.T.T. in "Knight Rider"); a comrade (Lt. Commander Data in *Star Trek: The Next Generation*); a conqueror/overlord (*The Matrix*); a dictator (*With Folded Hands*); an assassin (*Terminator*); a sentient race (*Battlestar Galactica/ Transformers*); an extension to human abilities (*Ghost in the Shell*); and the savior of the human race (R. Daneel Olivaw in the *Asimov's Robot Series*).

The reader can find further information about considered topics in articles [1–30].

Appendix 3
Current Supercomputers

A **supercomputer** is a computer that is at the frontline of current processing capacity, particularly speed of calculation. Supercomputers were introduced in the 1960s and were designed primarily by Seymour Cray at Control Data Corporation (CDC), which led the market into the 1970s until Cray left to form his own company, Cray Research. He then took over the supercomputer market with his new designs, holding the top spot in supercomputing for five years (1985–1990). In the 1980s, a large number of smaller competitors entered the market, in parallel to the creation of the minicomputer market a decade earlier, but many of these disappeared in the mid-1990s' "supercomputer market crash."

Today, supercomputers are typically one-of-a-kind custom designs produced by "traditional" companies such as Cray, IBM, and Hewlett–Packard, who purchased many of the 1980s companies to gain their experience. Since October 2010, the Tianhe-1A supercomputer has been the fastest in the world; it is located in China.

The term *supercomputer* itself is rather fluid, and today's supercomputer tends to become tomorrow's ordinary computer. CDC's early machines were simply very fast scalar processors, some 10 times the speed of the fastest machines offered by other companies. In the 1970s, most supercomputers were dedicated to running a vector processor, and many of the newer players developed their own such processors at a lower price to enter the market. The early and mid-1980s saw machines with a modest number of vector processors working in parallel to become the standard. Typical numbers of processors were in the range of four to sixteen. In the later 1980s and 1990s, attention turned from vector processors to massive parallel processing systems with thousands of "ordinary" CPUs, some being off-the-shelf units and others being custom designs. Today, parallel designs are based on "off-the–shelf" server-class microprocessors, such as the PowerPC, Opteron, or Xeon, and coprocessors like NVIDIA Tesla GPGPUs, AMD GPUs, IBM Cell, FPGAs. Most modern supercomputers are now highly tuned computer clusters using commodity processors combined with custom interconnects.

Relevant here is the distinction between capability computing and capacity computing [31], **Capability computing** is typically thought of as using the maximum computing power to solve a large problem in the shortest amount of time. Often a capability system is able to solve a problem of a size or complexity that no other computer can. **Capacity computing** in contrast is typically thought of as using efficient cost-effective computing power to solve somewhat large problems or many small problems or to prepare for a run on a capability system.

Figure A3.1 The Columbia Supercomputer at NASA's Advanced Supercomputing Facility at Ames Research Center.

In general, the speed of a supercomputer is measured in "FLOPS" (*FLoating Point Operations Per Second*), commonly used with an SI prefix such as tera-, combined into the shorthand "TFLOPS" (10^{12} FLOPS, pronounced *teraflops*), or peta-, combined into the shorthand "PFLOPS" (10^{15} FLOPS, pronounced *petaflops*). This measurement is based on a particular benchmark, which does LU (Logical Unit) decomposition of a large matrix. This mimics a class of real-world problems, but is significantly easier to compute than a majority of actual real-world problems.

"Petascale" supercomputers can process one quadrillion (10^{15}) (1000 trillion) FLOPS. Exascale is computing performance in the exaflops range. An exaflop is one quintillion (10^{18}) FLOPS (one million teraflops). The typical supercomputer is shown in Figure A3.1.

A3.1 Current Fastest Supercomputer System

Jack Dongarra has stated that the Tianhe-1A supercomputer in China at the National Supercomputing Center in Tianjin is 1.4 times as fast as the AMD Opteron-based Cray XT5 Jaguar at the Oak Ridge National Laboratory. According to Nvidia, Tianhe-1A has achieved a processing rate of 2.507 petaflops on the LINPACK benchmark. Tianhe-1A consists of 14,336 Intel Xeon CPUs and 7168 Nvidia Tesla M2050 GPUs with a new interconnect fabric of Chinese origin, reportedly twice the speed of InfiniBand. Tianhe-1A spans 103 cabinets, weighs 155 tons, and consumes 4.04 MW of electricity. The dethroned Cray XT5 Jaguar has a sustained processing rate of 1.759 PFLOPS.

Currently (July, 2011), Japan's K computer, built by Fujitsu in Kobe, Japan is the fastest (8.162 PFLOPS) in the world. It is three times faster than previous one to hold that title, the Tianhe-1A supercomputer located in China.

In February 2009, IBM also announced work on "Sequoia," which appears to be a 20 petaflops supercomputer. This will be equivalent to 2 million laptops (whereas

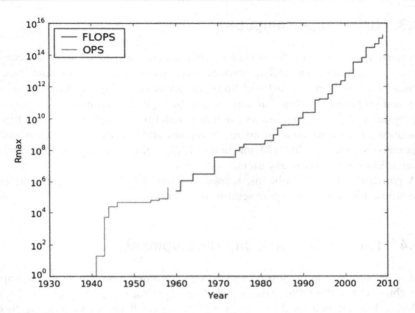

Figure A3.2 Fastest supercomputers: log speed vs. time.

Roadrunner is comparable to a mere 100,000 laptops). It is slated for deployment in late 2011. The Sequoia will be powered by 1.6 million crores (specific 45-nanometer chips in development) and 1.6 petabytes of memory. It will be housed in 96 refrigerators spanning roughly 3,000 square feet (280 m^2).

A3.2 Timeline of Supercomputers

Timeline of last supercomputers (Figure A3.2):

2004		70.72 TFLOPS	DoE/IBM Rochester, MN, USA
2005	IBM Blue Gene/L	136.8 TFLOPS	DoE/U.S. National
		280.6 TFLOPS	Nuclear Security
2007			Administration, Lawrence Livermore National Laboratory, CA, USA
2008	IBM Roadrunner	1.026 PFLOPS	DoE-Los Alamos
		1.105 PFLOPS	National Laboratory, NM, USA
2009	Cray Jaguar	1.759 PFLOPS	DoE-Oak Ridge National Laboratory, TN, USA
2010	Tianhe-1A	2.507 PFLOPS	National Supercomputing Center, Tianjin, China

A3.3   Using Supercomputers

Supercomputers are used for highly calculation-intensive tasks such as problems involving quantum mechanical physics, weather forecasting, climate research (including research into global warming), molecular modeling (computing the structures and properties of chemical compounds, biological macromolecules, polymers, and crystals), physical simulations (such as simulation of airplanes in wind tunnels, simulation of the detonation of nuclear weapons, and research into nuclear fusion), cryptanalysis, and the like. Major universities, military agencies, and scientific research laboratories are heavy users.

A particular class of problems, known as Grand Challenge problems, are problems whose full solutions require semi-infinite computing resources.

A3.4   Further Research and Development

IBM is developing the Cyclops64 architecture, intended to create a "supercomputer on a chip." Other PFLOPS projects include one by <u>Narendra Karmarkar</u> in India, a C-DAC effort targeted for 2010, and the Blue Waters Petascale Computing System funded by the NSF ($200 million) that is being built by the NCSA at the University of Illinois at Urbana-Champaign (slated to be completed by 2011).

In May 2008, a collaboration was announced between NASA, SGI, and Intel to build a 1 petaflops computer, Pleiades, in 2009, scaling up to 10 PFLOPs by 2012. Meanwhile, IBM is constructing a 20 PFLOPs supercomputer at Lawrence Livermore National Laboratory, named Sequoia, which is scheduled to go online in 2011.

Given the current speed of progress, supercomputers are projected to reach 1 exaflops (10^{18}) (one quintillion FLOPS) in 2019.

Erik P. DeBenedictis of Sandia National Laboratories theorizes that a zettaflops (10^{21}) (one sextillion FLOPS) computer is required to accomplish full weather modeling, which could cover a 2-week time span accurately. Such systems might be built around 2030.

References

The reader find the main author's (and about author) articles in http://Bolonkin.narod. ru) (see also: http://arxiv.org, http://www.scribd.com, http://archive.org, http://aiaa.org, http://www.km.ru, http://pravda.ru, http://n-t.ru, etc.), search term is "Bolonkin."

English

[1] Bolonkin, A.A., 1999. The twenty-first century: the advent of the non-biological civilization and the future of the human race. Journal Kybernetes Vol. 28 (No. 3), 325–334. MCB University Press, 0368-492X (English).

[2] Bolonkin, A.A., 2004. The twenty-first century—the beginning of human immortality. Journal Kybernetes Vol. 33 (No. 9/10), 1535–1542. Emerald Group Publishing Limited 0368=482X (English).

[3] Bolonkin, A.A., Human Immortality and Electronic Civilization, Lulu, USA, 1999. http://www.lulu.com, http://www.scribd.com/doc/24053302 (English), http://www.scribd.com/doc/24052811 (Russian), http://www.archive.org/details/ HumanImmortalityAndElectronicCivilization (English), http://www.archive.org/details/ HumanImmortalityAndElectronicCivilizationInEussian (Russian).

[4] Bolonkin, A.A., New Concepts, Ideas and Innovations in Aerospace, Technology and Human Sciences, NOVA, USA, 2007, 502 pp. http://www.scribd.com/doc/24057071 or http://www.archive.org/details/NewConceptsIfeasAndInnovationsInAerospaceTechnology AndHumanSciences.

[5] Bolonkin, A.A., The post-human civilization. Twenty first century: the end to biological mankind and occurrence of a post-human society, 1993. Basic articles are in Internet. http://Bolonkin.narod.ru.

[6] Bolonkin, A.A., Natural purpose of mankind is to become a God, 1999. Internet. http:// www.scribd.com/doc/26833526.

[7] Bolonkin, A.A., 2009. Human Immortality and Electronic Civilization. Publish America, Baltimore, MD. 145 pp.

[8] Bolonkin, A.A., Cathcard, R.B., Macro-Projects: Environment and Technology, NOVA, USA, 2008, 536 p. http://www.scribd.com/doc/24057930 or http://www.archive.org/ details/Macro-projectsEnvironmentsAndTechnologies.

[9] Bolonkin, A.A., 2006. Non Rocket Space Launch and Flight. Elsevier, 488 pp. http://www. scribd.com/doc/24056182 or http://www.archive.org/details/Non-rocketSpaceLaunchAndFlight.

[10] Bolonkin, A.A., 2010. New Technologies and Revolutionary Projects. Scribd, 324 pp. http:// www.scribd.com/doc/32744477 or http://www.archive.org/details/NewTechnologiesAnd RevolutionaryProjects.

[11] Bolonkin, A.A., Memories of Soviet political prisoner. Translation from Russian. Lulu, 1995, http://www.scribd.com/doc/24053855.

[12] Bolonkin, A.A., LIFE. SCIENCE. FUTURE (Biography notes, researches and innovations), Scribd, 2010, 208 pp. 16 Mb. http://www.scribd.com/doc/48229884, http://www.archive.org/details/Life.Science.Future.biographyNotesResearchesAndInnovations.

Russian

[13] А. Болонкин, «Жизнь. Наука. Будущее» (биографические очерки), Scribd, 2010, 286 pp. http://www.scribd.com/doc/45901785 or http://www.onlinedisk.ru/file/606043/, or http://www.archive.org/details/Life.science.futureinRussian....
[14] А. Болонкин, Бессмертие людей и электронная цивилизация (in Russian). 3-е издание, 2007. http://www.scribd.com/doc/24052811/.
[15] А. Болонкин, Природная цель Человечества—стать Богом. 2002. Internet. http://www.scribd.com/doc/26753118.
[16] А. Болонкин, Пост—человеческая цивилизация. XX1 век: Конец биологическому человечеству и возникновение пост-человеческого общества, 1993г, Основная статья в Internet. http://Bolonkin.narod.ru/.
[17] А. Болонкин, XX1 век—начало бессмертия людей, 1994г, Основная статья в Internet. http://Bolonkin.narod.ru/ (A. Bolonkin, 21 Century is beginning of human immortality).
[18] А. Болонкин, Поселим Бога в компьютер-интернетовскую сеть, 1998г. Internet. http://Bolonkin.narod.ru/ (A. Bolonkin, Settle God into computer-Internet net).
[19] А. Болонкин, Бессмертие становиться реальностью. Интервью Болонкина Б. Крутову, 1999г. Internet. http://Bolonkin.narod.ru/ (Bolonkin's interview).
[20] А. Болонкин, Наука, душа, рай и Высший Разум, 1999г, Основная статья в Internet. http://Bolonkin.narod.ru/ (A. Bolonkin, Science, soul, paradise, and artificial intelligence).
[21] А. Болонкин, Прорыв в бессмертие, 2002г, Основная статья в Internet. http://Bolonkin.narod.ru/ (A. Bolonkin, Breakthrough to immortality).
[22] А. Болонкин, Человеческое Бессмертие и Электронная Цивилизация, Lulu, USA, 1999. http://www.lulu.com (A. Bolonkin. Human immortality and electronic civilization).
[23] A.A. Bolonkin, Our children may be a last people generation, *Literary Newspaper*, 10/11/95, #41 (5572), Moscow, Russia (in Russian).
[24] A.A. Bolonkin, Stop the Earth. I step off. *People Newspaper*, September 1995, Minsk, Belorussia (in Russian).
[25] A.A. Bolonkin, End of Humanity, but not End of World, *New Russian Word*, 3/6/96, p. 14, New York, USA (in Russian).
[26] A.A. Bolonkin, Method of recording and saving of human soul for human immortality and installation for it. US PTO. Application US11/613, 380 filling 12/20/06, disclosure document No. 567484 on December 29, 2004.
[27] А. Болонкин, Записки советского политзаключенного. Lulu, 1991 (in Russian). 70 стр. http://www.scribd.com/doc/24053537, http://www.archive.org/details/MemoirsOfSovietPoliticalPrisonerinRussian.
[28] G. Igor, Price of immortality, Moscow, EKCMO, 2003, 480 pp. (Fantastic, in Russian).
[29] O.G. Pensky, Mathematical Models of Emotional Robots, Perm, 2010, 193 pp. (in English and Russian). http://arxiv.org/ftp/arxiv/papers/1011/1011.1841.pdf.
[30] O.G. Pensky, K.V. Chernikov, Fundamentals of Mathematical Theory of Emotional Robots (in English). Russia, PSU, 2010, http://www.scribd.com/doc/40640088/.
[31] Wikipedia. Some background material in this book is gathered from Wikipedia under the Creative Commons license. http://www.wikipedia.org.

Printed in the United States
By Bookmasters